ENERGY
The Facts of Life

ENERGY
THE FACTS OF LIFE

Andrew Buckley

A MARTIN BOOK

Martin Books
8 Market Passage, Cambridge CB2 3PF

First published 1979
© Andrew Buckley 1979
ISBN 0 85941 086 2

Advisory Editor: Ray Dafter

Conditions of sale
All rights reserved. No part of this publication may be reproduced, stored in a retrieval system or transmitted, in any form or by any means, electronic, mechanical, photocopying, recording or otherwise, without the prior permission of the copyright holder.

Printed and bound in Great Britain by Lowe & Brydone Printers Limited, Thetford, Norfolk.

Preface

The question of energy, its price and future availability, continues to attract widespread attention from newspapers and broadcasting companies the world over. Much space and time has been devoted to debating the scale of world energy supplies and the possibility of future shortage.

For us, however, the main interest in energy centres on petrol station forecourts and in the bills we receive for our everyday needs: for heating and lighting our homes, providing hot water, cooking food and running appliances.

But society's dependence on energy goes further. Without it we could undertake only primitive tasks and our leisure-time activities also would be severely limited. In short we are dependent on energy for our entire way of life. So when supplies are disrupted or questions are raised over future availability it is natural for us to become concerned. Indeed, we have only to turn our minds back five years to a time of scrambling in queues for two gallons of petrol and unlit homes and factories to remember the consequences of energy shortage. We surely cannot take energy for granted again.

Triggered by the large jump in oil prices, the prices for most other goods and services have also risen rapidly. Money incomes have accelerated strongly as we have tried to protect our standard of living: so much so that the impact of these fuel price increases have been considerably softened, if not totally negated.

Yet President Carter continues his battle to make American society more energy conscious, to cut out waste and charge more realistic prices. Growing funds are also being made available by governments throughout the world for the development of nuclear power and the new so-called 'alternative energy sources', certain in the knowledge that one day these will be called on to make a major contribution to our energy needs.

Most of our energy comes from the fossil sources: reserves of coal, oil and natural gas laid down over millions of years. Irreplaceable, they are being used up at a rapid and accelerating rate. They cannot last forever.

In Britain the position has become further confused with the rapid and successful development of our traditional energy sources. Expanding oil and natural gas supplies are complementing our prodigious coal production and well established nuclear programme. We are at the point of becoming energy self-sufficient, a fortunate position we expect to maintain for many years to come. Yet the government continues to urge us to 'save it'. To our individual cost we now see in our fuel bills a reflection of the true value of the resources we are using. And high prices are here to stay.

So what is our current energy position and what does the future hold in this area, so fundamental to society's development? In short what are the energy facts of life?

One aspect that has become clear is the virtual impossibility of predicting accurately the future availability of the various fuels or the likely developments in demand. Who, for example, would have predicted the events of 1973–74 even two years beforehand?

In formulating our national energy policy we will have to retain our flexibility: not to overcommit to the development of one fuel as opposed to another but to foster the development of the many alternative and renewable energy sources which promise so much for the future.

Nevertheless, as far as we are concerned as individual buyers of fuel we must not let the uncertainties of the future discourage us from examining the underlying trends and reappraising the way we look at and use energy in the light of them. It is our collective action now as consumers which can have a marked effect on the way we develop our energy resources in the years ahead.

A start to discussing these questions is made in this book. Ten 'Energy Facts of Life' form the chapter headings. They are:

Life depends on energy
High cost fuel is here to stay
The North Sea gives no room for complacency
We will all determine our energy future
Good housekeeping means lower fuel bills
Energy-saving makes a common-sense investment
Choosing appliances carefully pays off
Fuel costs need regular checking
Budgeting wisely avoids shocks
We must stay energy conscious

These 'facts' are not meant to be comprehensive. You will very probably think of more as you read on. Moreover the energy market is constantly changing. Prices of the various fuels change at different times during the year. Political developments may also have an immediate impact on the prices we pay and the availability of our fuels. This book, finalised in November 1978, attempts to be as up to date as possible. However, as events in December 1978 underlined, the energy situation is constantly changing and, in the light of developments in Iran and the outcome of the OPEC meeting in Abu Dhabi, a postscript has been added at the end of the book.

Andrew Buckley

Acknowledgements

In a book of this nature I find it almost impossible to cover the full range of people and organisations to whom thanks are owed. But firstly I must extend my special thanks to Ray Dafter of the *Financial Times* who agreed to act as the book's Advisory Editor. Through his extensive knowledge of the energy scene worldwide and his patient editing of the typescript he provided invaluable assistance and support during the lengthy task I set myself.

Thanks are also due to the many organisations who provided information and material in my researches, including the Department of Energy and the nationalised fuel industries (which in addition provided useful guidance at the manuscript stage). Here special mention should be made of the British Gas Corporation for the interest it showed during the preparation of the book and for making available its facilities at Marble Arch, including its extensive library, to a very grateful author. Support was also forthcoming from British Petroleum, for which again I am indebted.

Contents

	Preface	5
	Acknowledgements	9
1	**Life depends on energy**	15
	Progress demands energy	
	Our energy bank	
	Nuclear power	
	The renewable energy sources	
2	**High cost fuel is here to stay**	23
	The energy crisis in focus	
	The OPEC power base	
	Consumer power	
3	**The North Sea gives no room for complacency**	33
	Meeting our fuel requirements	
	Recent trends in consumption	
	North Sea oil	
	Natural gas	
	Coal	
	Electricity	
	Britain in the world market	
4	**We will all determine our energy future**	47
	Future energy supplies	
	Limiting our future needs	
5	**Good housekeeping means lower fuel bills**	57
	Cutting out waste	
	Coal prices	
	Oil prices	
	Gas prices	
	Electricity prices	
	Comparing prices and efficiencies	

6 Energy-saving makes a common-sense investment   71
    Roof insulation
    Cavity wall insulation
    Double glazing
    Controls
    Solar heating
    Heat pumps

7 Choosing appliances carefully pays off   85
    Central heating
    Room heaters
    Water heaters
    Other home requirements
    Six conclusions on fuels and systems

8 Fuel costs need regular checking   100
    How to read the meter
    Understanding your fuel bill
    Checking household consumption
    Questions and disputes on accounts
    Consumer councils

9 Budgeting wisely avoids shocks   109
    Budget payments
    Savings stamps
    Prepayment meters
    Budgeting for inflation

10 We must stay energy conscious   114

    Appendix : useful addresses   119

    Postscript   123

    Index   125

Illustrations and Tables

Table 1	World energy and oil consumption	16
Fig. 2	World fuel consumption since 1900	17
Table 3	Energy consumption in the principal Free World economies at the time of the energy crisis	26
Fig. 4	Domestic fuel price increases since 1972	28
Fig. 5	Trends in earnings and prices	29
Fig. 6	Movements of oil	30
Table 7	UK energy requirements 1977	34
Fig. 8	Current fuel consumption	36
Fig. 9	Crude oil gives a wide range of products	39
Fig. 10	North Sea oil and gas fields	49
Fig. 11	Our projected energy supply/demand balance	51
Fig. 12	Our energy profile: how much we need and use	53
Fig. 13	Fuel supplied to households	55
Fig. 14	Energy for use by householders after appliance losses	56
Fig. 15	Typical home fuel costs	59
Table 16	Estimated fuel efficiencies in home heating	69
Fig. 17	Typical areas of heat loss	73
Fig. 18	Insulating the roof space	75
Fig. 19	Low cost double glazing with plastic film	78
Fig. 20	Time clock controls for central heating systems	79
Fig. 21	A typical solar heating system	81
Fig. 22	How a heat pump works	83
Fig. 23	The growth in ownership of central heating systems	87
Table 24	Approximate costs for full central heating in a 3-bedroom semi	88

Table 25	Room heaters: typical purchase prices and running costs	92
Table 26	Water heaters: typical purchase prices and running costs	95
Fig. 27	Your gas bill	103
Fig. 28	Your electricity bill	104

1. Life depends on energy

We use energy in so many different ways and yet we hardly give a second thought to how it is produced or made available. Whether we are in the home or at work, or travelling between them, we are dependent on the use of energy to maintain, let alone improve, our way of life.

Our needs for energy do not stop when we switch off our car ignitions, heating systems or electrical appliances and lights. Energy is inherent in virtually everything we depend on: the food we eat, the clothes we wear and the jobs we do. Indeed, without energy life itself could not exist.

Progress demands energy

The amount we consume is dictated by our way of life. It has become the thermometer of our industrial progress and personal success. The larger the house we live in the more energy will be needed to manufacture the bricks and materials, and to provide the furniture and fittings to make it comfortable. And, of course, the bigger the house the greater will be our energy consumption to keep it heated adequately.

We in Britain use around eleven times more energy than the average African. Yet consumption in the United States and Canada is greater still – nearly two-and-a-half times European levels.

Industry is a prolific energy consumer. In Britain, for example, manufacturing companies take just under forty per cent of energy supplies. Commerce, too, is another important area, with its demands for heating and hot water in our shops, offices and public buildings.

As our requirements have grown more complex, therefore, and our lifestyle more demanding, so the rate of our energy consumption has accelerated. Today we use in one year what we used in a decade before the First World War.

Table 1. World energy and oil consumption
Developing countries use very little energy compared to the demands of the Western economies and the Soviet bloc. The rapid growth of industry in these countries will add even greater pressure to the demand for oil in the years ahead.

	Consumption per head of population, 1977	
(therms per year)	Primary energy	Oil
North America	3652	1687
USSR and E. Europe	1812	569
Australasia	1611	732
(United Kingdom)	(1610)	(700)
Western Europe	1478	806
Japan	1331	968
WORLD AVERAGE	696	309
Latin America	445	243
China	220	37
Africa	136	57
Asia (excluding Japan and China)	126	65

Source: British Petroleum

What are the prospects for the future? Left unchecked, and with the growing industrialisation of the Third World, consumption could double again by 1990 and perhaps yet again by the end of the century. But can our resources and technology meet such requirements? There is no simple answer. The world will not one day simply stop as the last gallon of oil or ton of coal is used. Equally, we cannot go on forever depending on the established sources – oil, gas and coal, the so-called 'fossil fuels' – to meet our ever-increasing demands.

Fig. 2. World fuel consumption since 1900
The world's massive and ever-expanding appetite for fuel is shown to have accelerated sharply since 1950. Against this background it really is not surprising that we are becoming increasingly concerned about a future energy shortage.

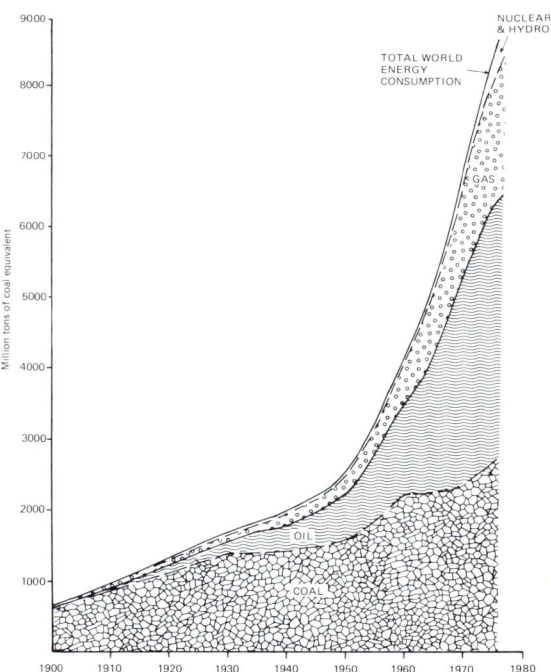

Our energy bank

Although no one knows just how abundant they are, these fossil fuels are, by their very nature, finite. Laid down as our energy bank in prehistoric times, they are available to us for use once and only once. We are currently using in twenty-four hours fuel reserves equivalent to a thousand years of prehistoric fossil development.

In Britain we are particularly fortunate in terms of fossil fuel reserves. Abundant coal supplies are now complemented by substantial North Sea oil and gas reserves. These will give us self-sufficiency in our energy requirements for some years ahead. But one day their supply will eventually fail to keep pace with our growing demands on them.

For our answer to the energy problem we must look in a number of directions. Firstly we must examine carefully our fossil fuel bank. Are we extracting as much as we can from known reserves? How many deposits of gas, oil and coal can we expect to find? And what are the timescales needed to develop alternative energy sources and nuclear power on a scale sufficient to maintain our energy needs in the future? We must also examine how we are putting to work our precious reserves. Are we using the right fuel for the job in hand? Is rapidly expanding energy consumption a prerequisite for further industrial progress and greater home comfort? Can we not take steps to eliminate waste and improve the efficiencies with which we put these fuels to work? These questions, of course, form much of the contents of this book, especially in regard to the use of energy in the home.

Nuclear power

We need also to examine the possibility of harnessing alternative energy sources. Nuclear generation of electricity has made only slow progress since the world's first large-scale station, our own Calder Hall, opened in 1956. It does, however, promise to meet much of our energy needs in the long term, offering, as at today's technology, the only viable alternative to fossil fuels. Nuclear power also brings with it novel environmental and security problems which have no ready-made solutions.

Less than two per cent of the world's energy requirements are currently met by electricity generated in nuclear stations. But this is a growing contribution, varying in importance by country. Both the USA and the Soviet

Union have sizeable nuclear industries, while Japan, France and Germany have important development programmes. In the United Kingdom nuclear power provided just over four per cent of our energy needs in 1977.

Nuclear fission, the process of splitting the atom, provides the heat source to drive the generating turbines. The atom is the basic form of matter and, on splitting, gives off large amounts of heat. This heat comes from the conversion of very small quantities of mass to energy.

In England the early Magnox stations are now being complemented by the AGRs – Advanced Gas Cooled Reactors. The AGRs are based on British technology but we have also decided to maintain an option to develop the American-based alternative, the PWR – Pressurised Water Reactor – in the early 1980s. We now have ten nuclear power stations providing nearly fourteen per cent of our electricity supplies. Three more stations are planned to become operational in the 1980s.

Much debate is under way on whether we can safely proceed from conventional techniques, using relatively large quantities of uranium, to the advanced 'breeder' processes which extract sixty times more energy from a given amount of uranium and also burn up some of the highly radioactive plutonium, a by-product from all nuclear fission reactors. Indeed, it has been calculated that the spent uranium extracted from existing stations would be sufficient to meet Britain's energy needs for the next 300 years if it was used to fuel fast breeder reactors.

We are only one of three countries, with France and the Soviet Union, to have built a prototype fast breeder reactor. Ours is in the north of Scotland and has been generating small quantities of electricity quite successfully since 1959. High-level discussions are currently taking place on the advisability of building a full-scale demonstration breeder reactor.

President Carter, despite his pressing energy problems, has deferred the development of such a scheme in the USA. There is widespread concern over these reactors, dependent as they are on the sensitive technologies of uranium enrichment, reprocessing and heavy water production, all fundamental to the manufacture of nuclear weapons. This concern is likely to hold back the development of fast breeders for some years to come.

Research efforts have also been directed to harnessing the prolific scope for energy supplies presented by nuclear fusion. This is the basic energy producing process of the sun, where, under conditions of intense heat and pressure, lightweight hydrogen atoms are packed together to form heavier helium atoms. In this process part of the matter is converted into energy.

Until very recently this process had been rendered impossible under controlled conditions by the requirement for exceptionally high temperatures, in excess of 44 million degrees centigrade. Scientists at Princeton University in the USA have, however, now produced temperatures of 60 million degrees centigrade using a magnetic field to contain the blazing gas. But this temperature has been achieved for just one-tenth of a second and work is continuing on the important second stage of the programme, to maintain these ultra-high temperatures for longer periods. Europe also has its nuclear fusion programme with research under way at Culham in Oxfordshire on the Common Market-sponsored JET project.

Whilst we are clearly many years away from nuclear fusion providing electricity on a commercial scale, its prospects have been greatly improved by the breakthrough at Princeton. Should these prospects be realised, it is claimed that fusion reactors would produce very little nuclear waste compared to the fission process. Furthermore, as deuterium, which is found in seawater, is one of the most suitable elements for nuclear fusion, this process could provide the world with all the energy it needs for many thousand million years.

Nevertheless, many problems remain to be overcome, both with nuclear fusion and with the fast breeder fission reactor. In the meantime, the nuclear contribution from conventional reactors will remain modest, dependent as they are on relatively large supplies of uranium.

The renewable energy sources

Stimulated by the energy crisis, scientists have also returned to examining the renewable sources of energy available to us. Many of these, such as wind and tidal power, along with the burning of wood, provided our ancestors with their first supplies of energy beyond their own muscles. And it is these renewable sources that may well hold the key for the world's energy supplies in the centuries to come.

All energy supplies originate from the sun. Its prolific and unending fusion ensures a continuing supply of solar radiation to our atmosphere. Some of this radiation is lost as it is reflected back into space by the upper atmosphere. But the rest is available to us, either reaching the earth's surface or being absorbed into the lower levels of the atmosphere.

Many schemes have been put forward for exploiting the alternative forms of the renewable energy sources that are available to us. An example of a source already harnessed is the electricity generated in hydro-electric plants. But in Britain this is small-scale and scope for further schemes is limited.

The Severn estuary presents one of the best sites in the world for producing tidal power. But the scheme, involving the construction of a massive barrage, would be expensive and thought likely, at current costings, to produce electricity at a greater cost than investing in further nuclear plants. Wave power, too, is under investigation and laboratory tests have shown the feasibility of harnessing this energy through a system of rocking concrete 'ducks'. Here again, problems need to be overcome. Current designs envisage large-scale structures which would pose difficulties for navigation. Wind power also offers potential but the size of plant required to generate sufficient power would also pose a serious environmental threat.

Geothermal energy is the name given to the heat which is constantly being thrust up from the earth's molten core to the cooler surface, as manifested in volcanoes and geysers. There is thought to be little prospect of harnessing such supplies in Britain. Overseas, however, the prospects are brighter: off the Californian coast, for example, pockets of hot brine have been located, which, if the problems of extraction and transmission could be overcome, would constitute a major new energy source.

But in terms of commercial development, solar heating itself probably offers the greatest short-term possibilities. Our weather and geographic position prohibit the direct collection of sunlight for generating electricity in large-scale solar furnaces, as tested in the southern states of America and elsewhere. Rather, solar energy offers a supplementary source to the heating systems of individual houses.

Sunlight is only one form of solar radiation; heating systems collect energy even if it is diffused by the clouds and atmosphere. With more energy falling on to British roofs through solar radiation than is used beneath them, the scope for greater development is apparent. But establishing solar heating systems is expensive, with costs taking ten years on average to recoup. We will be looking again at solar heating but the general conclusion here as regards all the renewable sources must be that they are unlikely to make any sizeable contribution to our energy requirements until the next century. The government's 'Energy Think Tank' has concluded that we can expect these sources to provide only six to eight per cent of our energy needs by the year 2000 and this contribution could only be arrived at with substantial investment in such projects as the Severn barrage scheme.

Therefore, we must turn back to the fossil fuels and combine their use with an effective campaign to cut out wastage and improve fuel efficiencies, and the development of a growing nuclear contribution, to provide our heat requirements for the remaining years of this century.

2. High cost fuel is here to stay

Speaking a few days before Christmas 1973, the Shah of Iran announced, on behalf of his fellow members in the Organisation of Petroleum Exporting Countries, that oil prices were to double for the second time within three months. The Free World was stunned. Could international banks and government treasuries, already under mounting pressure, cope with the massive transfer of wealth that would surely follow? How many more increases would there be? How could the consuming countries absorb them?

The energy crisis in focus

Today we are beginning to find the answers to these questions. The international monetary system was able to cope, albeit at one stage being on the point of collapse. Even so, the generation of these massive footloose funds remains a major problem in maintaining stable exchange rates. On the face of it we, and the other advanced economy countries, have weathered the storm. We are still here and trading, although at lower rates. But the impact on developing countries has also been harsh. Some have been helped by big increases in commodity prices which followed the trend set by oil but those less favourably placed have found life even tougher than before.

And what are the prospects for the advanced countries? Bedevilled by inflation and unemployment, stemming from the increases in oil prices, we have been stuck in the quagmire of the severest world recession since the 1930s.

But why was this collection of Middle Eastern and other developing countries able to inflict such a stranglehold on the powerful economies? It is a stranglehold retained today, keeping OPEC virtually the sole determinator of prices for all fuels throughout the Western world. The simple answer is that we have become overdependent on oil. And to make matters worse the developed countries – those with the greatest thirst for oil – cannot hope to meet the demand from their own reserves.

The remarkable increase in oil demand which has taken place since the 1930s has already been traced. Although it was produced in commercial quantities as long ago as 1859 it was not until the arrival of the internal combustion engine, offering personal mobility hitherto undreamt of, that the wider potential of this fuel was recognised.

At the same time oil was also beginning to replace coal for heating and the raising of steam. Although its widespread introduction to Britain for these· purposes was delayed by the Second World War, the 1950s and 1960s saw us making up for lost time. While British Rail switched to diesel-powered engines, British industry was finding the associated benefits of easier transportation, storage and use irresistible. During this period oil was cheap and abundant; seemingly only those with limited cash or foresight chose to ignore it.

In contrast, coal mining was being increasingly regarded as socially and economically undesirable. Requiring high manpower involvement, it was being shunned by the age of automation. By 1960 world production of coal had levelled off. In Britain's case coal production had been steadily declining, with few breaks, since reaching peak output before the First World War. In addition to losing its market on the railways, coal was also to lose out to oil in the manufacture of gas, before the arrival of North Sea supplies. In the face of rising energy demand, mines were closed and workforces reduced not only in Britain but in virtually all European countries. Indeed in some, such as Holland, coal industries were to disappear altogether.

The availability of cheap oil was such that not only was it masking the seriousness of our long-term energy problem, but it was also prohibiting the question being raised at all. It was also to have an important effect on another fuel, natural gas. Considerable quantities of this valuable fuel, produced with the oil, were being flared off as a waste product simply because at that time it was not considered economic to transport it to market.

Natural gas had been in use for many years in the USA for a wide range of heating applications. By the late 1950s smaller quantities were also in use in the Soviet Union, Canada, Italy and France. But, in Europe, it was not until 1959 with the discovery of the prolific Slochteren field in Holland that the possibilities of major natural gas supplies were envisaged. Dutch gas production quickly built up, not only meeting all the country's needs but also with some going for export. Sold at contract prices before the oil price explosion, these supplies are now seen to have been used too indiscriminately with little regard to the many premium benefits this fuel enjoys and bearing no relation to current oil prices. The so-called 'Dutch disease' stands as an important reminder in our own use of natural gas described in the next chapter.

Cheap oil also reduced the need for swift progress in the development of nuclear electricity supplies. So, as the Shah announced the redoubling of oil prices, energy consumption among OPEC's principal customers closed in 1973 in the unenviable position outlined in Table 3.

Oil was now the undisputed master of the energy scene and in most countries was meeting at least half of the total energy requirement.

The OPEC power base

Member countries of OPEC currently produce two-thirds of the Free World's oil. More importantly they are responsible for nearly ninety per cent of the oil made available on the world markets. They also control seventy per cent of published world oil reserves. OPEC's members

Table 3. Energy consumption in the principal Free World economies at the time of the energy crisis

In 1973 all the principal countries of the Free World were heavily dependent on oil. All, except the USA and Canada, were importing virtually all their requirements from OPEC countries.

1973 Energy Consumption Million tonnes of oil equivalent		Percentage			
	Oil	Gas	Coal	Nuclear and Hydro	
USA	1828	45	31	19	5
Japan	356	76	1	17	6
West Germany	272	55	10	32	3
UK	225	50	12	35	3
Canada	191	44	22	8	26
France	183	69	9	14	8
Italy	139	75	10	7	8
Netherlands	77	54	42	4	–
Australasia	69	51	6	36	7

Source: British Petroleum

come not only from the Middle East and North Africa but also include Nigeria, Venezuela, Ecuador and Indonesia. The organisation was established in 1960 at a time when American technology dominated oil exploration and the multinational companies, the so-called 'Seven Sisters', controlled refining and distribution.

Initially the role of OPEC was defensive, to safeguard the interests of member states from the multinational producers. The early rounds of price increases were designed primarily to protect member states from world inflation and not, as they were to become later, to form the main thrust to this inflation. But once successful the organisation moved on to the offensive. In addition to the price increases, oil production and refining facilities were nationalised with Saudi Arabia, Iran, Qatar, Kuwait, Libya and Iraq leading the way.

We are all conscious today of by just how much prices have risen. In Britain, and before North Sea oil began to complement our gas supplies, the scale of these increases placed great strain on our balance of payments with the resultant fall in the value of sterling. Now the combined importance of North Sea oil and gas is increasingly shielding us against the need for costly imports. Without them Britain's plight would be very serious indeed.

Fig. 4 charts the price rises faced by domestic customers for the four fuels since 1972. Whilst oil has led the way, it can be seen that the prices of the other fuels have also risen. The influence of OPEC extends throughout the range of alternative fuels. Indeed, the significance of the OPEC increases has penetrated throughout our Western economic system, resulting in higher production and distribution costs and supporting claims for greater wage increases. Fig. 5 shows that since 1973–74 fuel and power prices in Britain have moved ahead more quickly than the overall retail price index and in line with average earnings. OPEC actions, whilst not the root cause of our inflation problems, have greatly accentuated them, causing worldwide recession and entrenched unemployment.

The predominance of the Middle Eastern countries in determining OPEC pricing policy can be gauged from Fig. 6, which traces the principal flows of oil around the world. Saudi Arabia, Iran and Kuwait between them provide over half the organisation's production. Saudi Arabia is in a particularly strong position, being the largest producer and having the most prolific reserves. She is also able to vary her production very substantially at short notice. In December 1976, Saudi Arabia used this power and with the backing of the smaller United Arab Emirates set about raising production levels in order to keep world prices at no more than a five per cent ceiling about the previous level instead of allowing them to climb by ten per cent as favoured by the other members. This move was widely interpreted as part of a more general test of strength between the Saudis and the more militant OPEC members.

Fig. 4. Domestic fuel price increases since 1972

Led by OPEC increases for crude oil, the prices we pay for all our fuels have increased substantially since the energy crisis. But there are signs that increases over the next year or so should be less severe.

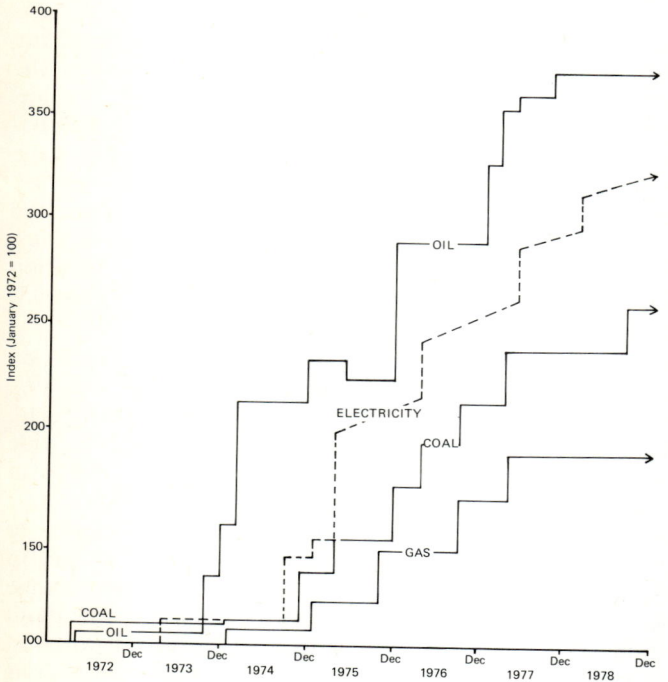

Source: Cambridge Information and Research Services

Fig. 5. Trends in earnings and prices
The published indices for fuel and light show that energy costs for the consumer have risen faster than the overall retail price index and in line with average earnings since 1973–74. OPEC increases, reinforced by the lifting of government controls on nationalised industry pricing, have kept energy prices moving ahead faster than inflation.

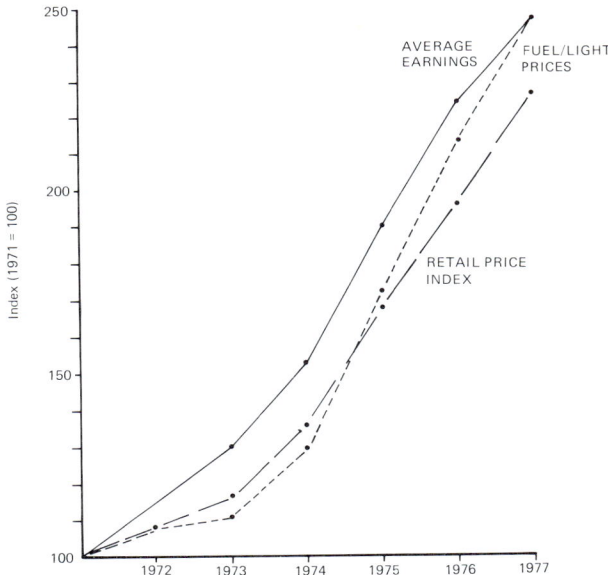

Source: HMSO *Annual Abstract of Statistics*

Indeed, when formulating price levels the organisation itself is as much influenced by political as economic pressures.

This makes the Western countries vulnerable not only on economic grounds but also dependent on the volatile politics of the Middle East. Nevertheless the West has regained some of the initiative seized by OPEC and it is to this question that we now turn.

Fig. 6. Movements of oil
The Free World is dependent on massive shipments of oil from the producing countries. The bulk of these supplies comes from the Middle East into Europe but imports into the United States are now increasing strongly.

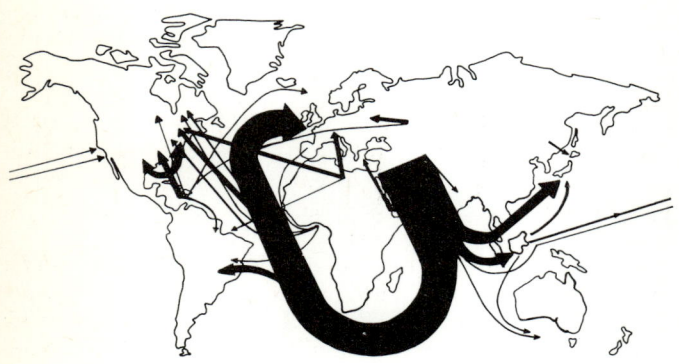

Source: British Petroleum

Consumer power

The compromise pricing agreement between Saudi Arabia and the other OPEC producers was finally reached in the summer of 1977. This agreement and the decision taken subsequently in December 1977 and June 1978 to freeze prices were forced on the organisation primarily because of the world slump in demand for oil.

For the first time since the energy crisis some of the initiative on pricing has been passed back to the consumer countries. Their efforts to limit OPEC imports, adopted

as part of the strategy of the International Energy Agency*
in 1974, were at last having some effect.

These efforts have been directed on three fronts:

to moderate the overall increase in energy demand through attention to and investment in conservation;

to stimulate development of alternative energy supplies such as the North Sea and Alaskan oilfields and the French and German nuclear programmes;

to promote the use of other fuels wherever possible, such as the conversion of electricity generating stations from oil to coal firing.

High energy prices have, in effect, helped consumer countries in their efforts to limit OPEC imports. As far as Britain is concerned, of course, it has provided the oil companies with a greater incentive to exploit the North Sea and the prospects are bright for further gas and oil discoveries.

Results from conservation efforts have not, however, proved so successful. The International Energy Agency found in 1977, during its first reappraisal since the energy crisis, that oil imports, particularly into the USA, were rising much faster than expected. One of the reasons for this, according to the Agency, was that expenditure on conservation had been held back by the world recession.

Without doubt it has been the recession which has influenced OPEC pricing the most. And with many authorities expecting this to continue until 1980 or thereabouts it is unlikely that further OPEC price increases on a major scale will be possible for some time ahead.

The danger for future pricing remains at the point when the world recovery gets under way. All OPEC members are

---

*The International Energy Agency was set up in February 1974 under the auspices of the Organisation for Economic Co-operation and Development (OECD). Membership covers most of the advanced countries of Western Europe, North America, Japan and Australasia.

committed to price increases on some scale or another. It is feared in some quarters that these increases could be very substantial as producers firstly repair the damage to their purchasing power accentuated by the falling value of the dollar and then strive to improve it.

The challenge facing consumer countries must be to push ahead with the development of additional fuel supplies and improve the use to which we put these fuels. These questions, forming the key to our future, are taken up again in Chapter 4. The alternative may well be a return to the sky-rocketing price increases of the last few years.

3. The North Sea gives no room for complacency

In our discussion of the world energy scene we noted the favourable position of the United Kingdom when compared with her industrial competitors. Not only were we less dependent than most on oil at the time of the Arab embargo, because of our available coal and natural gas, but also the move to high energy prices has assisted the development of our own offshore fuel supplies.

So successful have our oil exploration efforts proved that we now look forward to a period of oil self-sufficiency. North Sea exploration has also yielded further substantial gas reserves to add to our already important supplies which have been coming ashore since 1968.

To these must be added some of the most prolific coal reserves, per head of population, in the world. We also have our own nuclear technology developed over twenty years of power generation.

But politicians' claims that Britain is an island built on coal and surrounded by oil (and gas) must be treated with caution. We form only a tiny part of a fast-moving world energy market. Our energy self-sufficiency is assured for a relatively short period, by most current estimates of the order of twenty years. And even during this time the prices of our home-produced fuels will be governed by those prevailing in the world markets.

The offshore reserves identified so far suggest that we should be able to find a good deal more. The key to our future self-sufficiency, as discussed in the next chapter, depends very much on the continued development of our reserves and the improved uses to which we put them.

Meeting our fuel requirements

In 1977 the United Kingdom's energy bill topped

£16,000,000,000: a massive figure equivalent to a weekly expenditure of £5.50 per head of poulation.

In heat terms, this consumption was equivalent to 338 million tonnes of coal. Table 7 itemises this 'primary consumption' by the four fuels we used in tonnes of coal equivalent and in the other measures of bulk fuel supply, therms and tonnes of oil equivalent. (We will incidentally, be threading our way through the maze of fuel unit measurements and showing you how to compare them in Chapter 5).

Table 7. UK energy requirements 1977

We currently consume, in total energy terms, a little under half of the annual oil output of Saudi Arabia, the world's most prolific oil producer. Fortunately oil now is needed for just forty per cent of our energy requirements and all this is expected to be supplied from the North Sea by 1980.

	Consumption in 1977			
Fuel	Million tonnes of coal equiv.	Million tonnes of oil equiv.	Million therms	Percentage change on 1976
Coal	122·7	72·4	30,675	+ 0·6
Petroleum	136·6	80·3	34,150	+ 1·7
Natural Gas	62·8	36·9	15,700	+ 6·8
Nuclear/Hydro	16·3	9·6	4,075	+10·1
Total	338·4	199·0	84,600	+ 2·6

Source: Department of Energy

Not all this 'primary energy' is delivered to the end consumer. Some of our coal and oil is processed into 'secondary fuels' such as electricity and coke involving heat losses during the conversion process. For example, it takes three units of primary fuel, such as coal or oil, to produce one unit of electricity to the consumer.

The electricity supply authorities have taken steps to keep these losses to the minimum. Larger, more efficient generating stations are replacing the older and smaller units. Modifications have been made to the transmission system.

Furthermore, the enormous power station boilers can burn very low grade coal or oil which would probably find little use elsewhere. Power stations will always be limited in their efficiency but the growing use of their waste heat, for example in fish farming, could well cut down these losses further in the future.

These conversion losses are, however, the main reason why electricity is more expensive than alternative fuels. Nevertheless it fulfils a vital role in providing us with some of the energy we need. It is virtually irreplaceable for lighting and for powering machinery and some domestic appliances. Furthermore, as seen in Chapter 4, once delivered to the home or factory, electricity can be used more efficiently thus improving its cost effectiveness against other fuels.

Recent trends in consumption

Throughout the 1960s and early 1970s primary energy consumption in the UK increased every year at a rate linked to the rate of expansion in our national wealth – known to economists as the 'gross national product'. This was the period of cheap energy and industrialists and architects paid scant attention to the cost of fuel when erecting factories, town halls and office blocks. Our energy needs kept rising at a rate typically equivalent to three-quarters of the annual increase in GNP.

But the energy crisis brought a halt to this trend. Hard on its heels came the drive for energy conservation as householders and business consumers alike have been urged by government to 'save it'.

Demand for fuel has certainly moderated. But to what extent this has been a function of the conservation campaign or due to the higher prices or the general recession remains obscure.

Although demand has begun to pick up again, this is at a lower rate than that established before 1973. Primary energy consumption in 1978 is still likely to be below that recorded in 1973.

Fig. 8 shows how our current fuel consumption cake is cut in terms of how we use it and how much each supply industry provides.

Fig. 8. Current fuel consumption
Much of our primary fuel supply is processed into other fuels such as electricity and coke, involving conversion losses of around thirty per cent. In 1977 a little over fifty-eight billion (thousand million) therms was delivered to consumers.

The fuels we use
Oil remains the most important fuel meeting our transport and some of our heating requirements. But today each of the other fuels meets a significant part of our energy needs.

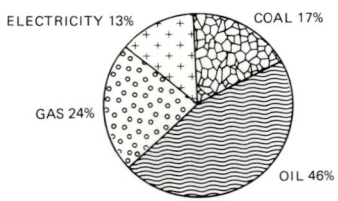

How they are used
Industry retains the greatest demand for fuel. But we also use nearly as much energy for transport as we do for heating our homes.

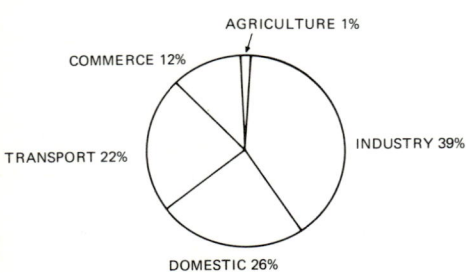

Consumption in the home
Our choice of fuels in the home varies considerably from the country's overall fuel use. Solid fuel remains an important choice but the advantages of gas and electricity, needing no storage and providing flexible and immediate heat, are reflected in their greater popularity in the home than within the total market.

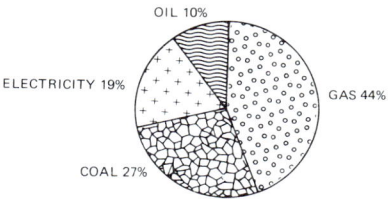

Source: Department of Energy

Industry, although its share of the total has been falling in recent years, remains the principal fuel consumer. Heat is required for all manner of purposes, for certain industrial processes, for maintaining working temperatures at the statutory 60° Fahrenheit and for welfare uses such as the provision of meals and hot water for washing.

In contrast fuel consumption in the commercial sector, which covers shops, offices and public services, is today above the 1973 level. Rising demand here has brought government action to improve conservation efforts. Energy use for transport, and particularly for road vehicles, has also continued to rise.

The government's first steps to limiting petrol consumption have included a major publicity campaign and the introduction of a new regulation requiring all new cars displayed in garage showrooms or forecourts to carry labels giving their official fuel consumption figures.

The figures compiled in tests, by the manufacturers but under the control of the Department of Energy, are for miles per gallon for urban driving and at a constant speed of 56 mph. There is an optional further test for consumption at 75 mph. A comprehensive list of test results is published half yearly by the Department of Energy and is available from HMSO bookshops.

And further action to limit petrol consumption cannot be ruled out. In the USA President Carter is moving inevitably to limiting engine sizes and similar moves involving tax concessions and penalties may well be introduced in Britain. Meanwhile the amount of energy required by households has remained surprisingly static in the face of new building programmes and improving levels of home heating. But it has been the move towards the use of more efficient fuels and systems which has brought about these improvements rather than the need to consume more fuel in the home.

There have also been important developments in the contributions each of the fuel supply industries have been making. Demand for coal from these final consumer markets (i.e. excluding that for electricity generation and coke production) has halved in the past ten years. Demand for oil continued to grow strongly until 1973 but fell back during the crisis and has only recently started to grow again.

Sales of natural gas have accelerated strongly in recent years, taking the industry's contribution to customer fuel requirements from eight per cent in 1968 to the current twenty-four per cent. Electricity has also increased its share of the fuel market.

We now look at each of these fuels and their current supply positions in a little more detail.

North Sea Oil

Since first supplies of North Sea oil were landed in 1975 production has built up rapidly and a level equivalent to self-sufficiency is expected by 1980. The development of these vigorously growing supplies with the promise of a lot more to come has transformed what was our ailing economy. Uppermost of all is the startling effect on our balance of payments, now moving into credit for the first time for many years. This has helped in halting the slide of sterling on foreign exchange markets – one of the reasons why our rate of inflation has become more moderate.

Production of crude oil from the North Sea topped a

Fig. 9. Crude oil gives a wide range of products
The basic process in oil refining takes place in the fractionating tower. Here the wide range of oil products are formed with each product meeting a different customer requirement.

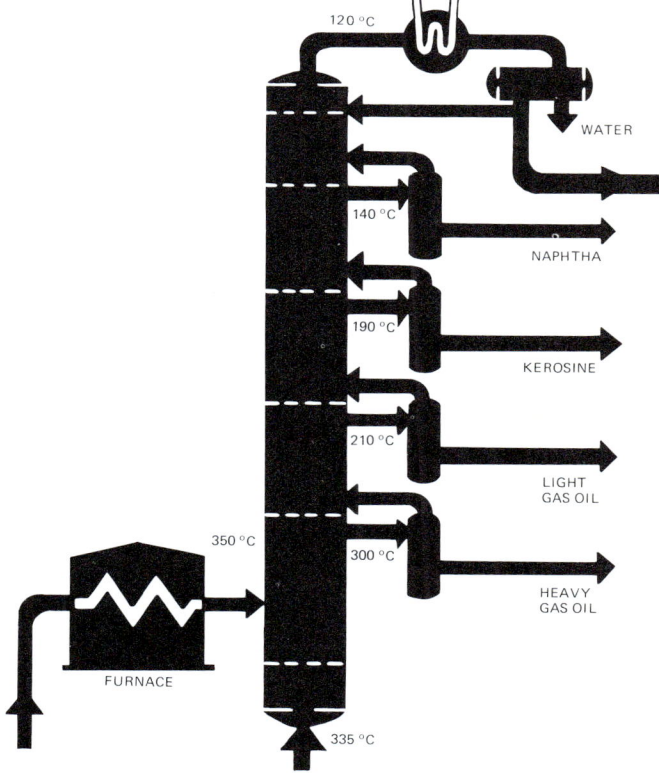

Source: British Petroleum

million barrels a day during May 1978, a level equivalent to three-fifths of our oil demand. In addition to the 13 North Sea fields on stream, a further six are under development there and a further 40 significant discoveries have been made, some of which will certainly be developed.

We will be examining the scale of the reserves these fields represent in the following chapter.

That oil meets a great deal of our energy requirements reflects the many products which are processed from 'crude', each having its own role to play. A simplified diagram of a refinery fractionating tower is shown in Fig. 9.

The crude oil is heated to a vapour and passed into this tower where the different products for processing are drawn off at separate levels. These levels are kept at different temperatures corresponding to those at which the various parts of the crude oil liquefy. Some remain as refinery gas which provides the liquid petroleum gases (LPG) supplied in bottles for household and caravan use or in bulk to tanks for industry and commerce. Next comes the important gasolene fraction, the basis of our different grades of petrol.

Kerosine is the basis for paraffin, used for portable home heaters and for the smaller central heating boilers and aircraft jet engines. Gas oil is used in industry and commerce for heating purposes with some also supplied for home central heating. Finally comes fuel oil, used in industry and power station boilers.

Natural Gas

The landing of natural gas supplies represented the first use of our North Sea energy reserves. Exploration in the southern waters off Norfolk, Lincolnshire and Yorkshire had followed the discovery of the prolific Dutch gas reserves. The availability of natural gas supplies posed a major challenge to the industry. Its heat content is about double that of the previous 'town' gas manufactured from coal and oil. It also possesses different burning characteristics.

But the decision was quickly taken to convert all gas-burning appliances to use this new fuel. After all, the switch to this richer gas would effectively double the capacity of the existing mains distribution system.

In April 1968 the conversion programme started. By the time it was completed, nearly ten years later, the homes of over thirteen million domestic customers had been 'converted' along with nearly four hundred thousand commercial premises and seventy-five thousand industrial establishments.

Supplies from six 'southern' North Sea fields are brought ashore through three east coast terminals and are distributed throughout the country by means of a newly constructed high pressure pipeline network. New supplies have recently started arriving from the Frigg field, which lies far to the north in the inhospitable waters off Scotland. Because of the field's exposed position, the high cost of exploration and the need to purchase some of the supplies from Norway, we are having to pay a lot more for Frigg gas than we do for that from the southern fields. Nevertheless, with conversion costs now paid for, the price of gas is likely to remain competitive and with the prospect of further supplies from the northerly Brent field and perhaps of other gas associated with oil, as well as gas from the Irish Sea, we are now assured of our own natural gas supplies for many years ahead.

Consumption of natural gas is expected to rise further at least until the early 1980s, by which time it could be supplying a third of our energy requirements. Existing knowledge of reserves would suggest that this higher level of consumption could then be maintained at least until the end of the century.

All this is a far cry from the industry's humble origins. Manufactured gas, produced by the carbonising of coal in retorts, was first used for street lighting in the nineteenth century. Although its uses were soon to expand to cover cooking, heating and industrial processing, coal remained the principal production source until the 1960s.

Then came its replacement by oil and the introduction of liquid natural gas (LNG) from the Sahara. This African gas was largely used in the production of manufactured gas at that time. The Gas Council pioneered the shipment of LNG in insulated bulk tankers, building a terminal on Canvey Island and laying the foundations of the new high pressure pipeline system.

The British Gas Corporation, as it is now styled, has also maintained its research efforts in producing gas from coal and oil. New techniques have been developed to produce this richer substitute natural gas (SNG) and the Americans, now realising the full value of gas supplies, have not been slow to seek our expertise. Furthermore, prospects exist for other methods of gas production using technologies so far untried. These include production from nuclear power using reactors to produce hydrogen, the farming of kelp and anaerobic digestion of waste. These technologies remain in early stages of development and warrant no further comment here. Nevertheless British Gas is confident that when the day eventually comes for natural gas production to go into decline it will have sufficient SNG facilities to meet customer requirements.

Coal

In marked contrast to natural gas, coal production reached its peak in 1913. By then it had given Britain a head start by fuelling the industrial revolution and supplies were being widely distributed for export. Production that year reached 287 million tons as against today's level of around 123 million tons.

In use since Roman times, coal gained a new application with the invention of the first steam engine in 1705. Four years later pig iron was smelted for the first time using coke and from then on Stephenson, Watt, Arkwright, Wedgwood, Telford, Darby, Bessemer and others were to make their contributions by the mid-nineteenth century, based on the use of this fuel.

Coal is mined from seams of varying thickness which

were formed over many millions of years from the decay of great forests. This plant life became compressed and fossilised over different periods producing soft and hard coals. The hard coals, such as anthracite, are the oldest; they contain the highest proportion of carbon and burn the cleanest.

Most of the National Coal Board's output is supplied directly to power stations, heavy industry and coking ovens. This last category requires special coking coals while the bulk heat uses of the first two are met by the smaller sizes of coal and coal slacks. The larger sizes are used for the domestic market and for meeting space heating loads in smaller industrial and commercial plants.

But, as already explained, recent decades have seen the decline of coal mining throughout the Western world. Coal simply lost out to oil.

With the breaking of the oil crisis it was natural to look again to coal, which has the most prolific reserves of our fossil fuels, and especially for bulk heat processing such as firing electric power station boilers. In Britain a new 'Plan for Coal' was established with the agreement of the government, unions and management. Now expected to cost £4,000 million this blueprint for the industry's modernisation involves an ambitious programme of pit development including the opening of a completely new and rich field at Selby.

This scale of investment reflects the industry's prolific reserves, as discussed in the following chapter. Coal is central to the government's long-term energy strategy and the National Coal Board is currently planning to raise annual production to 170 million tons by the year 2000.

Mining has always been a labour-intensive business though aided over the years by mechanised cutters, loaders and other equipment. Safety too has improved considerably. The Plan, and subsequent proposals which map out how the industry could develop to the end of the century, seek to improve productivity and the quality of these growing coal supplies.

Of equal importance to the industry's future well-being will be its ability to sell its products more widely again. Nearly eighty per cent of all supplies are used either for electricity generation or by the iron and steel industry.

It is generally recognised that the problem confronting all associated with the industry is how to keep it in reasonable shape until it is called on again to fulfil a dominant energy role. When this will be depends on many factors, including how successful we are at limiting our energy demands, how long our oil and gas supplies last and how quickly the nuclear programme develops.

Electricity

Electricity is the fourth form of energy we have at our disposal, although it cannot be regarded in the same light as the fossil fuels which provide heat and power through combustion.

Electrical energy becomes available by a flow of 'current' which in effect is a flow of charged atomic particles (electrons) through conductors. Metals are good conductors and the most convenient to use for transmission.

Current is made in two ways – by chemical change or by induction. Chemical production of charged electrons is the basis of batteries – used widely for small electrical systems such as torches, radios, etc. When a wire which is part of a circuit is moved through the force exerted by a magnet a flow of charged electrons (or current) passes through the wire and it is this principle (induction) which is used for large scale electricity generation. Steam turbines or other engines drive a generator with coils of wire spinning around magnets and current is induced which can then be fed through distribution cables for customers' use.

Most of our power stations use steam produced by burning fossil fuels — mainly coal, some oil, and in rare cases gas. At present seventy per cent of electricity is produced from coal-fired stations with the balance fairly evenly split between oil-fired and the nuclear stations. Gas-fired plants provide less than two per cent of electricity generated.

The major uses of electricity take one of two basic forms – either for driving motors (in which the process of generation is reversed to produce motion) or for converting to heat by being passed through a resistance, as in an electric fire.

Widespread use is made of electricity in all three consumer markets. During the 1960s and early 1970s demand for electricity grew rapidly. But the industry's short-term prospects changed dramatically with the energy crisis as it was forced to raise prices substantially in the face of paying more for coal and, particularly, for oil for its power stations.

This highlighted one of the major problems for the industry, that of estimating forward maximum demand, an exercise essential to keep the industry's programme of new power station construction in step. The levels of future generating plant orders, their balance between nuclear and new coal-fired stations and the choice of supply consortium have become questions of concern beyond the industry itself and ones where government has become involved for wider social and economic reasons.

The problems of developing a large scale nuclear power industry have already been mentioned in Chapter 1. Nevertheless it is expected that these will have to be overcome and nuclear power stations, including the controversial fast breeder reactor, will play a major role in meeting our long-term energy demand.

Britain in the world market

In common with the great majority of our competitors our fuel policy, what there is of it, is based on treating oil as the balancing fuel to make up any shortfall between our overall demand and the other fuels available. Of course this shortfall is substantial and we carry on using oil for much of our basic requirements.

We have already noted that the Free World's dependence

on imported oil allowed OPEC to exploit its monopolistic position. The escalating price of oil has driven up the prices of all other fuels and as competitive pressures grow for remaining fossil fuel supplies so prices will go higher.

The way we look on and use our own fuel resources cannot be isolated from this world position. Firstly the central role of international companies in the exploration of North Sea oil, themselves with refineries overseas and markets to satisfy, ensures that pressures will be maintained for some of our supplies to be exported, although the British government maintains an active interest in the scale of North Sea oil exports through its participation agreements with the producer companies.

Secondly we must charge ourselves world prices to ensure that we realise the full cost of the resources that we are depleting.

Realistic prices are the surest safeguard to maintaining the important conservation effort. Other countries, without such fuel benefits, are having to respond to changed energy prices. Our competitiors are becoming more energy efficient and we would do well to adopt a similar attitude.

For these reasons it would not seem sensible to directly subsidise the prices of domestic fuels in order to assist the low-income households. Rather, any assistance should be more effectively given through the established channels of social security and supplementary benefits.

Of course if world prices were to fall below the costs associated with the exploitation of our expensive North Sea oil and gas reserves then we would have a rather different problem on our hands. And this is possible since OPEC prices bear no resemblance to cost and could be cut very substantially in as short a time as they were raised.

But fortunately medium- and long-term pressure on energy demand, and specifically that for OPEC oil, would seem to rule this out.

It is in our interests, therefore, that we move into a period of energy self-sufficiency but at the same time carry on charging ourselves the going market rate.

4. We will all determine our energy future

Our energy bank is moving into credit. And just like our own bank balance we've got to work at it to keep it that way.

This chapter will be looking at how we, as a nation, can do this: firstly by ensuring that we extract the maximum possible value from our energy bank and secondly by putting these resources to their best possible use.

In this way we will postpone, for as long as possible, the so-called 'energy gap' when production of our fossil fuels will no longer be able to keep up with our energy demands.

Estimates of just when a world energy gap is likely to appear continue to vary widely. Assumptions need to be made on economic unknowns, such as the true extent of oil reserves, but views must also be taken on political issues.

Will Saudi Arabia be prepared to carry on supplying ever-increasing volumes of oil in the face of continued Middle East hostilities? How effectively can President Carter gain support for his energy conservation programme from the American public? These are the types of questions that need to be considered.

In the case of the UK our lack of dependence on imported energy supplies makes the picture a little clearer. Most experts are now predicting that we should have sufficient energy supplies to meet our needs until at least the end of the century. After that the picture becomes less clear, with official projections of energy supplies incorporating, perhaps optimistically, a large-scale contribution from nuclear power.

It is, of course, the case with all forecasts that the further ahead they cover, the less reliable they turn out to be. With this cautionary note let us examine our prospects firstly for supplies and then for demand.

Future energy supplies

The principal North Sea oil and gas fields currently producing or under development are mapped out in Fig. 10 together with the growing network of undersea pipelines and land terminals.

By the time these oilfields reach peak production, sometime around the mid-1980s, dependent on continued progress in the face of hostile weather conditions, we will be producing each year getting on for one and a half to two times our current oil consumption. Already substantial quantities of oil have been earmarked for export to Germany and large quantities of associated liquid petroleum gases for North America.

The picture also looks bright for natural gas. As previously noted, production from existing fields, including Frigg, together with the large quantities of gas associated with the Brent oilfield, should support the industry's further expansion into the 1980s. By 1982, or thereabouts, British Gas could be supplying half as much again as it currently does.

The results of an in-depth study into the feasibility of harnessing smaller reserves of gas associated with the other North Sea oil fields have now been presented to the government. Although such a full-scale scheme does not appear to be economically viable, for the moment at least it is likely that, initially, gas from some oilfields close to existing gas lines into the St. Fergus gas terminal will be brought into these pipelines through short spurs. British Gas has also been active in exploring for gas in other coastal waters. Drilling in the Irish Sea has resulted in a commercial find in Morecambe Bay, a field expected to prove as big as some of the finds in the southern part of the North Sea.

Further oil exploration activity has also been under way. In addition to the fourteen producing fields there have

Fig. 10. North Sea oil and gas fields
Development of our North Sea oil and gas fields has been swift and successful. By 1980 we will be energy self-sufficient and sending supplies for export.

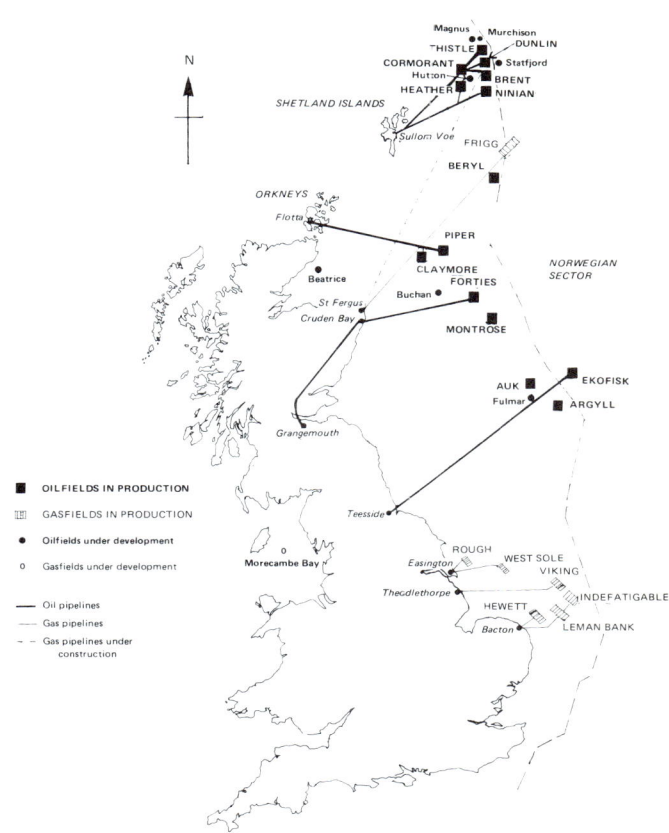

been numerous other finds, some of which are going into development. Interest remains in the areas to the east of Scotland and around the Shetlands.

Oil has been struck to the west of the Shetlands – indeed the prospects look to be exciting, but so far exploratory drilling in the Celtic Sea has been disappointing. Further to the south the Western Approaches could well contain sizeable oil and gas fields. Britain's offshore exploration and production venture is still in its early days and even the most conservative of estimates incorporate provisions for further supplies to come from so far undiscovered fields. It is also common practice for secondary exploration to yield more finds which were missed in the initial exploratory stages.

It is perhaps also worth noting as we pass that current recovery rates of oil are around twenty-five to thirty per cent of the total oil in place. This low proportion is not confined to the British fields but is common to oil exploitation throughout the world.

Oil reaches the surface through the natural pressures built up in the earth's crust. Advanced recovery techniques are now being developed, for example by pumping down chemicals or gas to maintain pressures. These hold the promise of extracting a lot more oil from each known reservoir.

The French Petroleum Institute has estimated that by raising recovery rates from twenty-five to forty per cent world oil reserves could last for over a hundred years at current consumption rates. A lot more work is needed here and clearly there are going to be no short-term gains. Nevertheless as prices go higher so we can expect improvements in recovery rates.

This bright picture of our North Sea oil and gas reserves is traced out in Fig. 11 together with our coal and nuclear potentials. These estimates cannot hold any specific degree of accuracy. Instead they should signify the way in which our reserves are developed.

Existing knowledge tells us that we have extractable

Fig. 11. Our projected energy supply/demand balance
Growing supplies of North Sea oil and gas, added to our coal and nuclear electricity output, should ensure that we have more than enough fuel to meet our energy needs into the 1990s. The government will need to decide what proportion of our surplus supplies we should export.

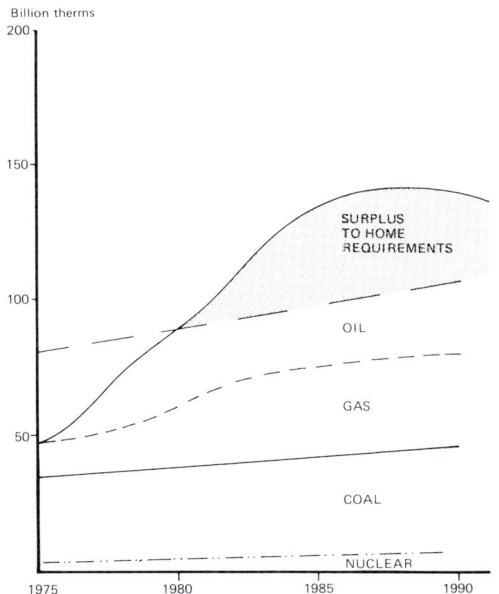

Source: Cambridge Information and Research Services

coal reserves to support current consumption levels for at least the next three hundred years. As previously noted, an ambitious investment programme is in full swing with costly plans under consideration. The very scale of these reserves ensures that this fuel will play a key role as oil and gas fields near depletion. And if our land-based deposits prove insufficient it is widely believed that further vast deposits lie buried under the North Sea. This should present a suitable challenge for the technology of the twenty-first and twenty-second centuries.

Long-term energy plans also place heavy reliance on the development of a large-scale nuclear power industry. As previously noted, such reliance demands under current technology the use of breeder reactors with all their consequential implications. Our immediate goal of making full use of fossil fuel reserves can hopefully delay such agonising decisions if not eliminate them altogether.

Limiting our future needs

Against this rather optimistic view of our energy resource bank it is perhaps surprising that we need to consider how we can moderate our future requirements.

But there can be no guarantee as far as future supply projections are concerned. There will always remain the possibility of major accidents, such as the one on the Ekofisk field, seriously disrupting production as well as causing pollution. Some fields may develop faults in production with today's estimates of their reserves in hindsight proving over-optimistic.

Nor can the UK's energy future be isolated from that of Europe and the rest of the world. Shortages elsewhere are also bound to affect us. There are indeed many uncertainties facing our future energy supplies.

What is perhaps more certain is that the price of all fuels will continue to rise. Within the context of our own pockets then, if not for any altruistic motives, we are likely to be looking for ways of saving fuel. Most of the remaining chapters in this book offer advice on how we can do just that. But before we get to these let us look and see what overall scope we have.

Fig. 12 gives an overall impression of the amount of energy we lose in the processes involved in obtaining the heat and energy we need.

Firstly there are the losses involved in producing the types of fuel we need from those we have available. We have already noted the scale of these conversion losses in, for example, the generation of electricity from coal and oil. Other examples are the use of coal in the manufacture

Fig. 12. Our energy profile: how much we need and use
Much of the fuel we need today is lost before and during the
time we use it. Losses occur firstly on conversion to other fuels,
such as the use of coal in generating electricity (from Stages
1 to 2). Substantial losses are involved when we actually use
the fuel (from Stages 3 to 4). Although we will never be able
to use our fuel with total efficiency the chart nevertheless
underlines the great scope we have to cut down on fuel losses.

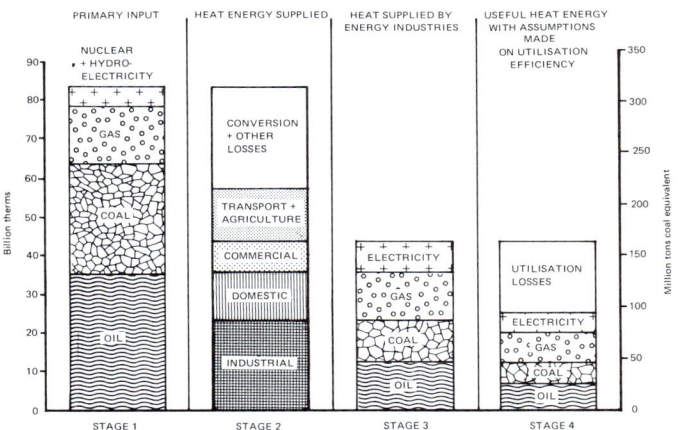

Source: Cambridge Information and Research Services

of coke and the need to use some oil for refinery purposes.
We can see in Fig. 12 that from a primary input of a little
over 84 billion therms, conversion and distribution losses
claim 30 per cent, allowing 58 billion therms to be
delivered to the end consumer.

To these 'conversion losses' inherent in producing the
types of fuel we require we must also add those to losses
associated with appliances or plants used by the final

consumer. Unlike the conversion losses, which can be readily measured, these 'utilisation losses' are more difficult to assess and can only be estimated since they involve the way appliances and plant are used as well as their tested efficiency.

The losses are calculated here on the basis of broad assumptions of the overall efficiency of each fuel within each of the three heat-energy markets. These figures are open to dispute but the purpose of this exercise is to provide a general understanding of the loss levels involved.

The scale of the loss is clear from Fig. 12: excluding the use of energy for transport and agriculture the amount supplied to the three heat markets – industry, commerce and domestic – is less than 45 billion therms. But of this amount around forty per cent is thought to be lost in use.

So whilst it is impossible in practice to eliminate utilisation losses completely there is clearly ample scope for improvement.

An interesting example of where such an improvement has been taking place is the domestic market (Fig. 13). Whilst it is obvious that the use of coal has been declining to the benefit of the other three industries, especially gas, the most remarkable feature of the illustration is the overall static level of consumption at around 14 to 15 billion therms over the entire period.

Yet during this period many new houses have been built and the general standard of household heating has improved substantially. How can this have been the case when overall consumption has remained static?

The answer is contained in Fig. 14, which depicts a very different situation. Measured in terms of the amount of energy consumed after appliance losses have been deducted, it shows a steadily increasing total. We are, as a nation, now getting more useful heat out of the same amount of delivered fuel.

It should be borne in mind however that, although we have achieved this important improvement in the way we use fuel in our homes, we can't expect increased efficiency

Fig. 13. Fuel supplied to households

The total amount of fuel supplied each year to domestic customers has remained surprisingly static since 1960. Within the total it can be seen that the demand for gas has risen considerably, compensating for the fall in coal deliveries.

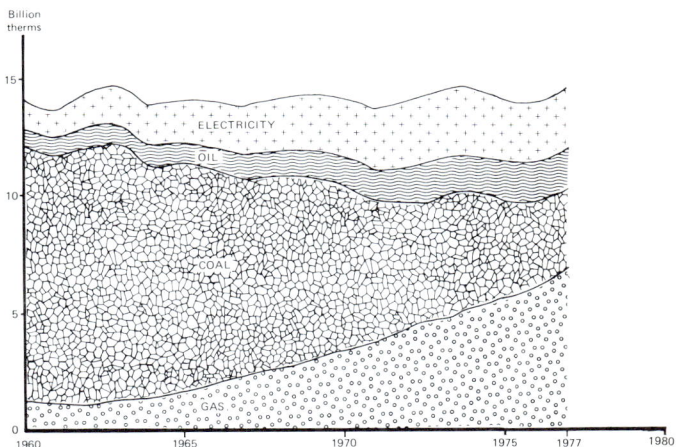

Source: Department of Energy

always to compensate for future increases in demand. Other methods of energy saving must also be used if we are to prevent a rise in the total fuel supplied to homes in future years.

But clearly attention to improving appliance and fuel efficiencies is an important issue, one which spells hope for energy use on a worldwide scale. Making better use of fuel not only in the home but at work means that we can continue to improve our economic activity and further improve our standard of living up to a point without necessarily involving a similar increase in energy consumption.

Fig. 14. Energy for use by householders after appliance losses
In contrast to the previous graph we see the total amount of useful heat available rising. But these figures have been calculated after allowances for appliance losses and hence show that improvements in heating standards are stemming from the use of more efficient fuels and appliances rather than greater fuel consumption.

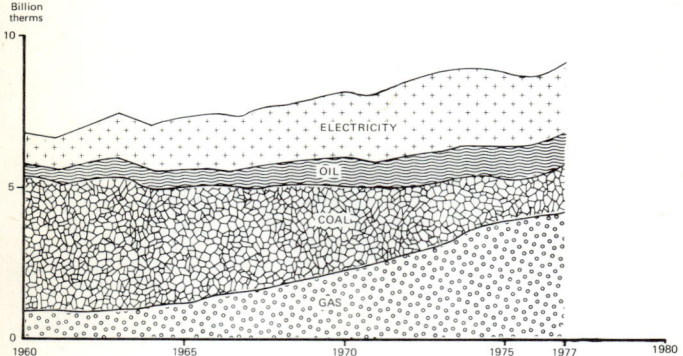

Source: Cambridge Information and Research Services

Cutting down energy waste in conversion, delivery and use, combined with the general drive to 'save it', should result in overall savings of twenty per cent or more on demand levels in the 1990s. This is a very important saving and one which, pricing issues apart, should encourage us all to improve our efforts in fuel management.

5. Good housekeeping means lower fuel bills

Faced with the recent steep rises in fuel prices, and the likelihood of continuing increases, attention to how we use and purchase our fuels now pays handsomely.

We will therefore be looking at just what form this attention should take over the next four chapters. Here we will be covering the things that we can all do which, while costing us virtually nothing in terms of cash, can bring significant savings.

Action can be taken in two areas. Firstly we can take elementary, but highly effective, measures to retain the heat that we are already using in our homes and ensure we are not using it wastefully.

Secondly, it is important to understand the basic characteristics and relative prices of our fuels. Each has its own units of measurement and each contains varying amounts of 'useful heat', as defined in the previous chapter.

All fuels are widely available. Oil and coal are available throughout the country from a network of merchants. Electricity is supplied to virtually every household; some nineteen and a half million domestic customers. Gas is not quite so widely available, especially in country districts. There are 14 million domestic customers for this fuel.

Later we will be looking at the prices and characteristics of these fuels and showing how their units of measurement can be compared and how their efficiencies in use correspond in our different energy requirements. But firstly we deal with the fundamentals of good housekeeping.

Cutting out waste

Fuel bills are now approaching an average of £200 per household. If you have some form of central heating installed (and about half of all homes have), or are keeping a family home reasonably warm throughout with individual appliances, then your combined fuel bills could well be around £400 a year and sometimes much more.

If this is the case a realistic saving, say of twenty per cent, is going to give you £80 in cash within twelve months. A £100 saving is certainly not impossible. At today's tax rates that is equivalent to a wage rise of over £150 a year, a target well worth achieving.

Once you have the money you may consider spending some of it on insulation and other methods of energy saving to enable you to save even more fuel. We will be looking at these options in the next chapter.

But first we have to create that initial saving of fuel and money. Fig. 15 shows just where in the average home our fuel consumption goes. It is obvious that we must attack the home heating bill first.

Draught-proofing of doors, windows and floorboards will save you fuel and make you feel more comfortable. Materials are cheap and easy to use. They are readily available from do-it-yourself shops and you will normally recover sensible levels of expenditure within six months through lower heating bills.

All exterior doors and windows should be draught-proofed and hollow wooden floors covered whilst keeping the space underneath the floorboards well ventilated. Shrinking floorboards and ill-fitting skirting boards form prime entrance points for cold air. Pack old newspaper, torn into strips, into these crevices to halt the flow of cold air.

It has been estimated that some homes are so draughty that it is equivalent to having ten bricks taken out of the living room wall!

Of course draught-proofing can be taken too far, rendering rooms odorous and poorly ventilated, and even unsafe. This is particularly the case if they contain appliances burning oil, gas or solid fuel. Adequate amounts of air are needed for these to burn properly and efficiently.

Fig. 15. Typical home fuel costs
Out of every £1 spent on fuel in the home nearly two-thirds goes to heating rooms. Cutting down any unnecessary waste associated with room heating must therefore be the first step. But don't forget your hot water supplies. Simple measures taken here can bring big dividends.

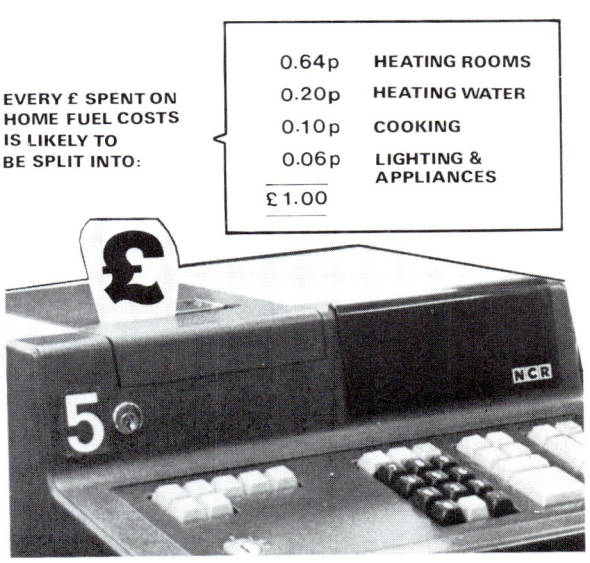

EVERY £ SPENT ON HOME FUEL COSTS IS LIKELY TO BE SPLIT INTO:

0.64p	HEATING ROOMS
0.20p	HEATING WATER
0.10p	COOKING
0.06p	LIGHTING & APPLIANCES
£1.00	

Source: Department of Energy

Products of combustion also need to be disposed of to avoid dangerous levels of carbon monoxide building up. Flues and vents should be regularly checked.

Heat losses through windows are much greater than through walls. One option, to install double glazing, requires substantial expenditure, especially if you have it fitted by a specialist contractor. We will therefore hold over discussion on this, and the important question of loft insulation, until the next chapter.

The use of heavy lined curtains will also help in cutting back heat loss through the windows. Make sure the curtains fall below the window ledge and also if you have a radiator beneath the window it is important to keep the curtain behind this, allowing the heat from it to circulate into the room and not go out through the window. And remember also not to place heavy furniture in front of radiators but allow the heat to circulate freely.

It is also well worth while fitting, by means of sticky tape, cooking foil or similar materials behind radiators. This foil has the effect of reflecting heat which otherwise, laboratory tests have shown, is partially lost through the brickwork. This is particularly effective on older houses with solid walls where the scale of heat loss is greater than in cavity wall construction.

Excessive air changes in a room also waste fuel. Whilst only one and a half changes of air are necessary every hour to keep rooms smelling sweet, it has been found in some houses using open fires that up to six air changes take place every hour. Fitting a throat restrictor will reduce unnecessary excessive ventilation of an open fire by up to fifty per cent. But remember with all fuels care must be taken to maintain adequate flueing. Expert advice will bring the best results whilst ensuring safety standards are preserved.

If you have a central heating system big savings can also be made by the careful use of its controls and thermostat. You will probably get a ten per cent fuel saving if you lower your heating system thermostat by as little as three degrees Fahrenheit. Many homes are running at too high a temperature during the 'working day' with windows open for proper ventilation and fuel wasted.

Regulating air temperatures throughout the house will also pay dividends. Systems are geared, through radiator sizes and positioning, to provide varying temperatures in different rooms, but turn the radiator in the spare bedroom right down – not off completely or you will get condensation forming.

Make use of your radiator controls throughout the day. You cannot be in the living room and bedroom at the same time so don't heat both continuously. More sensitive temperature controls specifying room temperatures can also be fitted to radiators and can prove helpful.

Give attention also to the time-controls linked to the system. (If you have an old system a time clock may not have been fitted as a standard. If this is the case get a qualified electrician to fit one, the investment will prove well worth while).

Time controls should be adjusted to give you the heat when you want it – not before or after. If the house is empty all day make sure it is not reaching its temperature peak as you shut the front door on your way to work.

Equally, if you or a member of your family remains at home during the day you may find the intermittent use of a fire preferable than keeping the central heating system running. This should cost a lot less and may also prove more comfortable being able to give a higher level of localised heat. Most kitchens, it should be remembered, generate sufficient heat on their own account.

But if you move over to an individual fire bear in mind that because of standing charges small quantities of different fuels bring relatively higher overall cost. In these circumstances it is best to select a fire which uses the same fuel as the central heating system.

Attention to your hot water requirements can also pay off. If you do not have your storage tank insulated by now then you have probably wasted over £150 in fuel bills since the energy crisis. The cost of a three-inch jacket is around £5. It takes very little time to fit. And if you have a jacket already fitted it is worth checking that it is not one with a one-inch thickness. If so, replacing it with a three-inch one will bring significant saving.

Techniques for saving water, found necessary during the drought of 1976, will naturally save you fuel when they affect the hot tap. You get three showers for the price of one bath. Do not run the system too hot – reducing the water temperature a few degrees brings much greater savings through reducing heat losses throughout the system.

Kitchens do not offer such substantial scope for energy savings. And much of the scope which is available stems from the use of common sense. Use appliances only when you really need to. Keep fridge and freezer doors closed. Boil only the amount of water you need and not the amount that the kettle can take.

Implementing these simple measures will involve little effort or cost yet should bring significant fuel savings. It is important also to understand a little about the fuels you use and their prices to make sure you get the most for your money. We now look in turn at the four fuels available.

Coal prices

Coal and other solid fuels have been traditionally measured and sold by weight – hundredweights and tons. From early 1978 these measures are being superseded by the metric units, kilograms and tonnes and these will fully replace imperial weights by mid-1979.

Sacks will normally weigh 50 kilograms or about 110 pounds. The metric tonne, which contains 1000 kilograms, is a little lighter than our old ton. (There are 2205 pounds to the metric tonne as against the old 2240 pounds to the ton.)

Different grades of solid fuel mainly reflect the varying levels of impurity content, e.g. ash. The heat content or calorific value (c.v.) varies with these different grades.

In the case of coal calorific value is expressed in Btu's – British thermal units – normally per ton. A Btu is the amount of heat required to raise one pound of water by one degree Fahrenheit. There are 100,000 Btu's in a therm, the standard measurement for gas, mentioned later. It is also expressed in the European measurement of gigajoules per tonne (GJ/t).

Average coal provides 265 therms per ton (27·52 GJ/t). House coal has a higher calorific value around 290 therms per ton (30·11 GJ/t) with anthracite as high as 321 therms per ton (33·33 GJ/t).

Domestic coal is distributed through a network of fuel merchants, some of whom also sell oil products. Prices tend to rise with distance from the coalfield leading to the highest prices in the South West, South East and East Anglia.

House coal is only suitable for use in open fires in areas outside smoke-control zones, or for burning in appliances specially exempted by the Department of the Environment. These appliances, of which a few are currently available and more will be developed, burn the cheapest coal without emitting substantial quantities of smoke. The current average price for Group 2 coal is around £42 per tonne rising to over £50 in some areas. Higher grade anthracite nuts, suitable for slow burning in closed stoves, boilers and ranges, are currently upwards of £59 per tonne.

Prices have increased regularly during the past few years and the Coal Board raised those for domestic fuels by 10 per cent from November 1978.

Sunbrite is the brand name given to domestic coke, again for use in boilers, stoves and cookers. The price is between £47 and £66 per tonne. Homefire, Phurnacite, Rexco and Dry Steam are other brand names for higher quality fuels. Prices range between £50 and £75 per tonne.

Oil prices

Petroleum products have traditionally been measured in both weight (tonnes for large quantities) and in volume (gallons). Here too Britain is harmonising with continental measures with heating oil already costed out and delivered in litres. (There are just over four and a half litres to the gallon).

Each oil product has its specialised uses and each varies in weight and heat content. The lightest products are the liquid petroleum gases, propane and butane. They also have the highest heat contents at around 470 therms to the tonne. LPG is used for portable appliances in homes and caravans and as an alternative for cooking in rural areas not served by mains gas.

Towards the centre of the heat content range come petrol (445 therms to the tonne), kerosine vaporising oil (435 therms) and gas oil (431 therms). Domestic oil-fired boilers normally use kerosine whilst gas oil is sometimes used in the larger systems but more typically in offices and factories.

The heaviest products are the fuel oils used in industry for raising large amounts of heat. Their higher impurity contents keep the calorific values down (to around 406 therms per tonne).

The heating values of the oil products most commonly used in the home and for transport are:

petrol	445 therms per tonne or 1.51 therms per gallon
kerosine evaporising oil for heating systems	435 therms per tonne or 1.62 therms per gallon (0.36 therms per litre)
gas oil used in larger heating systems	431 therms per tonne or 1.64 therms per gallon (0.36 therms per litre)

Oil for central heating is obtained by delivery from fuel merchants. Scheduled prices from the large companies were introduced some years ago to reflect the differences in costs of delivery from the refineries. These so-called inner and outer zones show differences of about 0·1 pence per litre. But the increases in overall price over the past three years have been so large that they are no longer significant.

With a surplus of oil products at the high prices brought about by the OPEC increases, as detailed in Chapter 2, merchants have been competing for business and differences at present of around 0·5 pence per litre for domestic fuel have been commonplace.

Current market prices are around $8\frac{1}{2}$ to 9 pence per litre for deliveries of 500 gallons for home heating oil.

Kerosine is normally purchased in smaller quantities at around 10 to 10½ pence per litre. As explained in Chapter 9, however, the purchasing price per litre of oil supplied may not be the sole criterion for selecting a distributor. Many run schemes incorporating servicing arrangements, a necessity when using oil, but at very different rates. Others may offer extended credit terms again at differing costs. All these factors need to be taken into account when considering various deals.

Price increases are subject in part to decisions taken by the OPEC producers. The expected increase in January 1979 is likely to form the first of a series of regular moderate adjustments.

Gas prices

Gas is measured in volume but charged for in therms. In Britain cubic feet are the standard units on gas meters. The calorific value in Btu's per cubic foot is specified by the supply authority and checked by government testing.

Manufactured gas at 500 Btu's per cubic foot has been replaced by natural gas at just over 1,000 Btu's per cubic foot. Gas meters are tested for accuracy within legal limits and stamped by government department before use.

Bills are charged in therms (100,000 Btu's) and the price per therm is specified in the published tariffs. As with other fuels burned this gives the heat input into the appliance – output is dependent on the appliance efficiency.

Tariff structures have been simplified in recent years. Today domestic customers are charged on a single tariff unless a prepayment meter is installed.

Over eighty per cent of gas customers use the credit meter system. Meters are read quarterly and accounts are normally presented within a few days. Payment is therefore made after consumption.

The Domestic Credit Tariff incorporates a small standing charge currently between £1.50 and £2 per quarter depending on the gas region. This standing charge is levied regardless of use and represents the customer's contribution to the cost of the supply and metering. The initial block of consumption, currently covering the first 52 therms per quarter, is charged at a relatively high price (19.3 to 22.8 pence per therm depending on the region) with all further gas consumed being charged at the lower level of 15.3 pence per therm. Credit customers using gas for cooking and little else will therefore pay a higher rate per therm than those making wider use of gas for central heating where most of it will be charged at the lower rate.

There is no standing charge attached to the Domestic Prepayment Tariff but British Gas recoups the contribution to the cost of meters and supply by levying a higher rate on the first 39 therms per quarter. This highest rate – currently between 23.8 and 29.3 pence per therm, depending on the gas region – produces an annual sum about equal to the standing charge for credit customers. Quarterly consumption above 39 therms is normally at a similar rate to the first block of the credit tariff but the further reduction applied to credit customers is not applicable.

The Prepayment Tariff is therefore suitable only for small consumptions. Not only is it more expensive for any sizeable use of gas but the collecting coin boxes have only limited capacity. There is also the disadvantage that pilot lights and other burners have to be relit once coin payment measures have been exhausted.

Coin meters normally accept 5, 10 or 50 pence pieces and are emptied at seventeen-week intervals. The meter mechanism is set to provide an exact quantity of gas for coins inserted related to a relevant setting. At the time the meter is emptied the published selling price of gas is obtained and additional monies refunded to the customer.

The range of prices per therm quoted reflects the differences in tariff levels among the twelve gas regions. Each region produces its own tariff leaflets and these are available on application to your local showroom.

British Gas has announced that tariffs will remain frozen until April 1979 subject to the general rate of inflation being kept under control. This means that tariffs will have remained unchanged for two years.

Electricity prices

Electricity is sold by the unit. One unit is consumed by using electricity at the rate of one kilowatt (kW) – equal to 1,000 watts – for one hour. This means, for example, that one unit will provide ten continuous hours of lighting from a 100 watt light bulb or keep a 1kW fire operating for an hour. 1kW is equal to 3,412 Btu's – see page 62.

Bills are calculated by multiplying the units consumed by the appropriate price per unit in the published tariffs. Like gas, details on current tariff rates are available in leaflet form from each area board.

Prepayment meters have also been continued by the electricity boards and they will be prepared to install one where they consider it is safe and practical to do so. This would not however be acceptable if the meter is in a shared hallway or outside meter cabinet, where the risk of theft is considered too great.

Customers with prepayment meters are charged the same unit prices as those on the Standard Domestic Rate but are surcharged for the cost of provision and collection. This surcharge is currently 7 pence per week.

There are two main domestic credit tariffs: the Standard Domestic Rate and White Meter (Domestic Night and Day (Rate), which offers customers cheaper rate supplies during the night period, when the cost of supplying electricity is lowest.

The Standard Domestic Rate incorporates a standing charge of between £2.50 and £3.50 per quarter and units are charged at a rate between 2.5 and 2.7 pence dependent on electricity board area. There is a supplement added for fuel cost adjustment.

The White Meter Tariff carries slightly higher standing charges at £3.50 to £5.00 per quarter with night rate units charged at about half the rate for units on the Standard Domestic Tariff. But consumers on the White Meter Tariff do pay slightly more than the Standard Domestic Tariff rate for their daytime consumption, between 2.5 and 2.9 pence per unit.

Electricity Boards offer alternative Domestic Night and Day Rates depending on the length of the night period. All boards in England and Wales are now offering a cheaper tariff, Economy 7, for householders choosing a seven hour night period.

Customers with electric storage heating are usually charged on the Restricted Flow or White Meter Tariffs. But some on the Standard Domestic Rate and without storage heating could also benefit by changing over to the Economy 7. Those who do not use more than about twenty per cent of their electricity at night will not benefit from the White Meter Tariff.

All electricity unit rates today incorporate a fuel cost adjustment (FCA). The FCA was introduced to reflect unexpected changes in the price paid by the Central Electricity Generating Board for the coal and oil needed to produce the electricity. The workings of the FCA are discussed further in Chapter 8.

Tariff revisions have been taking place annually over the past few years – at which time the FCA surcharges may be incorporated into the new tariffs. The last general revision occurred on 1st April 1978.

Comparing prices and efficiencies

We have already noted that fossil fuels supplied to consumers are subject to sizeable losses on use. These utilisation losses in the domestic market are estimated to account for forty-two per cent of the heat supplied.

This is because gas, solid fuel and oil release their energy through combustion, i.e. by being burnt in air. The waste heat merely escapes into the atmosphere. The use of electricity does not involve such losses.

But unlike conversion losses, where it is a more straightforward matter say of measuring the amount of coal needed to generate a given output of electricity, utilisation losses can only be broadly estimated. They vary not only by fuel and appliance but also of course by how efficiently the appliance is working and used by the customer. The importance of regular servicing of appliances cannot be

Table 16. Estimated fuel efficiencies in home heating

The efficiency with which your appliance uses fuel will have a marked bearing on the price you pay. Efficiency rates vary depending on the fuel used, the application to which it is put and the performance of the equipment. The figures given below are typical of results achieved when modern appliances are maintained in good condition after being correctly installed.

Process	Fuel	Efficiency (%)
Fires	Electricity	100[1]
	Gas	55
	Solid fuel	40[2]
Storage heaters	Electricity	90[1]
Water heaters	Electricity	85[1]
	Gas	50
Oil heaters (flueless)	Oil	90
Central heating	Electricity	90[1]
	Gas	70
	Oil	70
	Solid fuel	65[3]

[1] Electricity is converted into heat at 100 per cent efficiency but as storage heaters produce heat throughout 24 hours, and during times when it is not required, overall efficiency rates are considered lower. Similarly heat losses from pipes and the tank need to be taken into account for water heating.
[2] Efficiency improves to up to 65 per cent in open fires with high output back boilers.
[3] A back boiler for hot water only operates at 45 per cent efficiency.

overstressed and is discussed separately in the next chapter.

Table 16 makes some attempt to measure the potential efficiency of the principal fuels in their main heat uses. The high efficiency in use of electricity is apparent, offsetting to an extent the high conversion losses in generation at the power station.

Moves toward efficiency labelling of appliances as a conservation measure are under consideration – to complement the recently announced miles per gallon display of test figures for new motor cars.

But comparing overall costs of fuels and appliances in use is not just a straightforward matter based on application tests and fuel prices. Other factors too must be taken into account – and it is more appropriate to compare the total overall costs of providing the complete service for average requirements. This we will be doing in Chapter 7.

6. Energy-saving makes a common-sense investment

Implementing the elementary measures described in the last chapter should bring down your fuel bills substantially, so much so that you should be encouraged to consider more costly forms of energy saving.

You will certainly not find a lack of things to spend your money on. Since the energy crisis firms have sprung up everywhere, offering a wide range of products and services sold to a greater or lesser extent under the mantle of 'energy saving'. But beware, some of these are costly to purchase and install.

You should ask yourself whether from an energy-saving standpoint they really are worth it. Or will the savings that are achieved be so small that it will take many years to recover such expenditure?

These are the questions we will be asking in this chapter in relation to the main forms of energy saving investments such as:

roof insulation;
cavity wall insulation;
double glazing;
heat controls;
solar heating;
heat pumps.

Whilst some measures won't cost much, such as fitting a jacket to the hot water tank or draught excluders to doors and windows, other schemes will require a great deal of money. So when we consider these we must understand whether they will bring sufficient fuel savings to justify the investment. In short, how quickly will we get our money back?

Remember that 64 pence in every pound spent on fuel in the home typically goes for heating our rooms. Once the hot water tank is lagged, which must be the top priority, our attention should turn to keeping as much heat as possible inside our living rooms.

It has been estimated that less than two million of the nation's 20 million homes are well insulated. Another million have some form of insulation but the rest have little or none at all. British houses are amongst the worst insulated in Europe.

Energy savings of 50 per cent or more are possible in a typical inter-war semi-detached house with cavity walls by methods of insulation, draught proofing and controlling ventilation. The use of some form of insulation is an obvious step, but where should we start – in the roof or with the walls? Or perhaps double glazing?

Fig. 17 shows the areas where heat typically escapes from the home – both before and after insulation measures have been taken. In the first instance we can see that over one-third of our heat losses go out through the walls. The other principal escape route is through the roof, although the floor and doors are also important.

Once measures have been taken these losses can be cut by well over 50 per cent. The figures to the right of the illustration show what proportion of the original heat input to the house is now escaping. These are important because they show different rates of saving. Roofs and walls should clearly be our areas of first interest. Attention here should cut losses by 40 per cent.

Roof insulation

Enough loft insulation, to a minimum thickness of 3 in. (75 mm), for a semi-detached house will cost around £35–£40. If you had no insulation before, this should reduce your heating bills by at least £25 a year, and probably more. So you will get your money back within 18 months and thereafter the gain will be yours forever – that is, providing you don't lose the advantage of the insulation by keeping the home at a higher temperature than before.

Less than a quarter of Britain's homes have the mini-

Fig. 17. Typical areas of heat loss
Insulating your house properly can cut the amount of escaping heat by over a half. The figures for 'before' and 'after' — left and right below — show roofs and walls as the areas where best results are obtained.

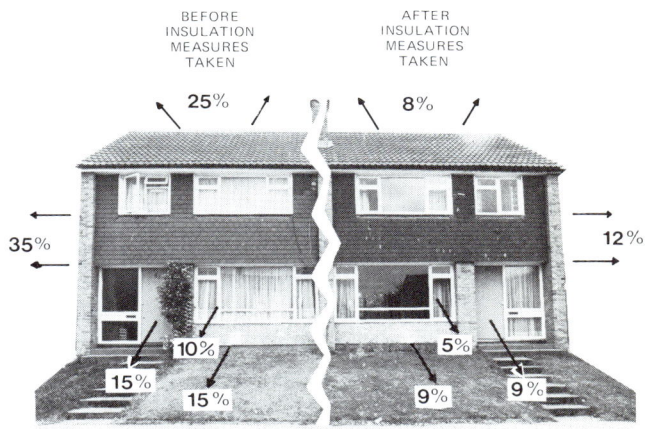

Source: National Cavity Insulation Association

mum 3 in. thickness of insulation. So unless you have recently installed it yourself, you may well find the existing layer thinner — that is if you have any at all. It's well worth checking. Upgrading from 1 in. to 3 in. thickness will cost about £25 and will bring savings in your fuel bills sufficient to cover this investment within three years, assuming you keep your upstairs temperatures down to the correct levels — see page 60.

Official recognition of this waste, costing upwards of £75 million a year in Britain alone, has come in recent announcements to improve standards in council-owned property and to provide grants to private householders to meet part of the cost. In a new scheme to be administered by local authorities private householders will be able to claim up to two-thirds of the cost of installing roof insulation and jacketing hot water tanks to a maximum of £50 per house.

Those eligible for this assistance include home owners, landlords and tenants, except those of a local authority or housing association, but only where the loft is without any form of insulation. The grant will not be forthcoming if you have already put some insulation in your loft, even though it may be below the standard, or if insulation is only required for just lagging the tank and pipes.

Applications for assistance should be made at your local council offices. The council will require an assurance that your house it not already insulated and may check this before approving your application. Once acceptance is granted the work can be carried out and the receipts presented for partial reimbursement.

With this added incentive there really can be no excuse for delaying insulating the loft. As Fig. 18 shows, this is a simple do-it-yourself exercise where all you need is the material, normally glass fibre or mineral wool, and a little time. Remember if you use granular infill insulation you will need a 4 in. (100 mm) thickness to obtain the same result: that is, unless you have a flat roof, where the job is much more difficult and one that needs expert attention. Indeed, you'll probably find it a better proposition to insulate other parts of the house instead.

Cavity wall insulation

Although insulating the roof space should take first priority our illustration (Fig. 17) did show that the area of greatest heat loss is in fact through the walls. With most modern houses this can be tackled quite easily.

Much depends on whether your house is constructed with exterior cavity walls. The great majority of houses built since the war have such cavities as it was found that the two walls, with a space or cavity between, was one of

Fig. 18. Insulating the roof space
A simple do-it-yourself job, laying insulation in the loft will cut heat losses and save you money. Here glass fibre is being unrolled between the joists. You need at least a 3 in. thickness or a 4 in. layer if you use granular infill.

When you're in the loft make sure the pipes are properly lagged. Two methods are illustrated here. Remember that once you've insulated the roof it will be very much colder in the space above.

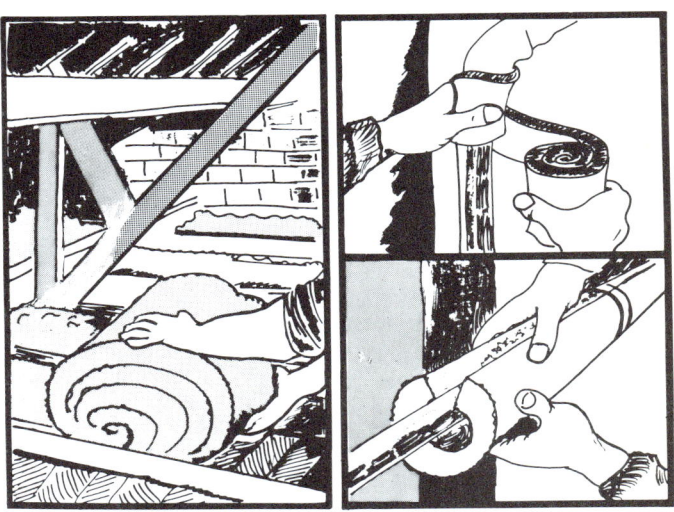

the best ways of avoiding damp spreading to the interior. Some houses dating as far back as the 1920s also have cavity walls.

If your house has solid walls then the problem is greater. You could consider lining the inside walls with insulation board, although there is a risk of condensation forming. Alternatively, insulating cladding could be fixed to the outside. But these jobs are not straightforward and you will probably have to seek advice from a local builder.

In contrast, insulating cavity walls is fairly straightforward, although here again you will need to enlist the services of a specialist contractor. The contractor will inject suitable insulation material, probably urethane foam or mineral fibre, into the cavity through holes drilled in the external walls. It will normally take him about a day and as he works from the external walls the operation should not involve any disruption or mess inside the house.

There are many firms offering this service and it is important to ensure that the job is done satisfactorily. Account needs to be taken of the condition of the external walls, their construction and exposure to weather. The walls were designed with the cavity to avoid damp. Whilst under normal conditions injecting insulation will not affect this, it is important to seek skilled advice. It's also wise to have the operation of all flues which pass through the wall checked after the work has been done, as a blocked flue will be dangerous. If you have any doubts about the effectiveness of a flue, always call in a competent heating installer to check it.

Many contractors will guarantee their work for 20 or 30 years, issuing a certificate to that effect. But as with all guarantees look carefully at the small print and at the reliability of the company giving it. When you come to sell your house this will obviously prove valuable.

Costs will vary according to the size and construction of the house but they are likely to run out at between £100 and £250. On this basis it may take up to five years before your reductions in fuel bills allow you to recoup this investment. After that of course you will continue to benefit from the savings. And if you come to sell the property then it should be worth that little bit more.

But before you go ahead, remember that in England and Wales cavity wall filling is subject to building regulations. Although these have been relaxed to promote its use in private homes it will still be necessary for the installer to notify the local council. If the council considers that the general relaxation does not apply in your case, possibly because the work is not being done in accordance with a certificate issued by the Agrément Board (a board assessing products and techniques used in the building industry), then a formal application will need to be made. In Scotland formal approval may not be needed but it is worth while notifying the building control officer before work is started.

Double glazing

Heat losses through windows are much greater in any given area than through the same expanse of external wall. It has been calculated that losses through windows are six times those through a similar area of insulated wall.

Nevertheless, window areas of houses are normally small and account for only 10 per cent of heat losses in an uninsulated house (Fig. 17). Moreover, even after attention, only half of that relatively small proportion will have been saved.

Installing double glazing is therefore unlikely to prove attractive looked at from a purely 'energy saving' viewpoint. Whilst you may have good reasons for going ahead with double glazing to reduce noise or eliminate cold spots the cost of installing it fully throughout the house could take as long as 30 years to recover through lower heating bills.

Do-it-yourself enthusiasts will obviously fare better in terms of cost, and there are many systems on the market. But from an energy-saving standpoint it is essential to select a system with a cavity width between the window and the inner pane of at least $\frac{3}{4}$ inch. If the gap is just $\frac{1}{4}$ inch it will prove only half as effective as those systems with the $\frac{3}{4}$-inch space.

For those prepared to spend a little time, benefits will be gained from putting up plastic film as a temporary measure during the winter. The technique described in Fig. 19 will prove particularly effective on cold spots, perhaps north-facing windows. And the cost of plastic, at around £2 for 12 square feet, is reasonable.

Fig. 19. Low cost double glazing with plastic film

A cheap way to eliminate cold spots, without going to the expense of full double glazing, is to use plastic film. The film is cut to the required size and fixed to the inside of the frame with double-sided tape. Whilst this forms an effective barrier to heat losses, it does have disadvantages. Being less rigid than glass, the film is likely to be quite noticeable. And when summer comes you will have to take it down, refixing with new tape for the following winter.

Controls

One of the surest ways of cutting fuel bills is of course to use the controls of your system wisely. Basically there are three types:

manual taps and switches;
time controls;
thermostats.

Central heating systems of both the hot water (radiator) and ducted warm air types incorporate manual controls for varying room temperatures throughout the house. Keep the radiators turned down in rooms not in regular use. Turn them down in the bedroom during the day and open them out again in the evening. A moment spent on this will pay handsomely in keeping down the boiler's consumption.

Most central heating systems incorporate an automatic time control. The typical example in Fig. 20 incorporates

Fig. 20. Time clock controls for central heating systems
The time clock is the control centre of a central heating system. The operating arms on the clock face enable the boiler to be switched on and off, usually twice a day. Having set these to the times you want, the boiler will then operate automatically.

To save fuel keep the running time of the boiler to the minimum. Start-up could be an hour before you normally get up: programme it to go off about half an hour before you usually go out. Set it to come on about half an hour before you get home and to go off half an hour before you usually go to bed.

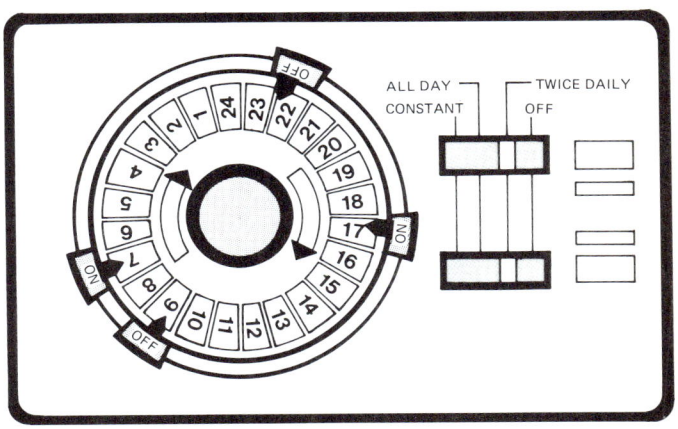

a twice on–twice off routine, allowing two periods of boiler operation within 24 hours. The settings are easily adjustable to allow for longer periods of heat at weekends and in the winter and also for periods of shutdown when the home is not occupied.

If your system does not include a timer, you can have it installed for around £30. With costs normally recovered within 12 months the use of an automatic time control should be regarded as essential to the efficient running of every central heating system.

Equally important is the use of air temperature thermostats for shutting down the boiler when the required temperatures are reached – typically 68°F in the living room and 65°F in the hall. Again, no central heating system should be without one.

Individual thermostats can also be fitted to radiators. At a cost of around £12 each they can keep rooms at varying temperatures throughout the house. But although they offer the capability of 'fine-tuning' heating levels, similar results can be achieved merely by adjusting radiator levels by hand.

Solar heating

Spurred on by the energy crisis and the growing concern of the future 'fossil fuel' shortage, interest has grown in harnessing solar energy as a potential source for heating domestic water. More than a dozen manufacturers of solar heating systems were represented at a recent energy exhibition in London. As interest develops so inevitably come the complaints of sharp trading practice and exaggerated sales claims. Great care must be taken in understanding just what these systems are capable of achieving.

The first point to realise is that, whatever the system, it can only act, due to the confines of our climate, in conjunction with a back-up system using a conventional fuel.

A typical solar water heating system is sketched out in Fig. 21. The core is the collector panel, which is essentially

Fig. 21. A typical solar heating system
The direct rays of the sun and the radiation scattered by the clouds and haze are harnessed by the blackened collector plate. Heat is then transferred through an exchanger to the hot water tank.

Although solar systems cannot provide a totally independent heating source for domestic requirements, used in conjunction with an auxiliary heating system they can bring about substantial fuel savings.

The collector plate is covered by glass to transmit the incoming radiation and to obstruct the outgoing reradiation. The plate is fixed over the existing roof involving little alteration to the tiles or slates.

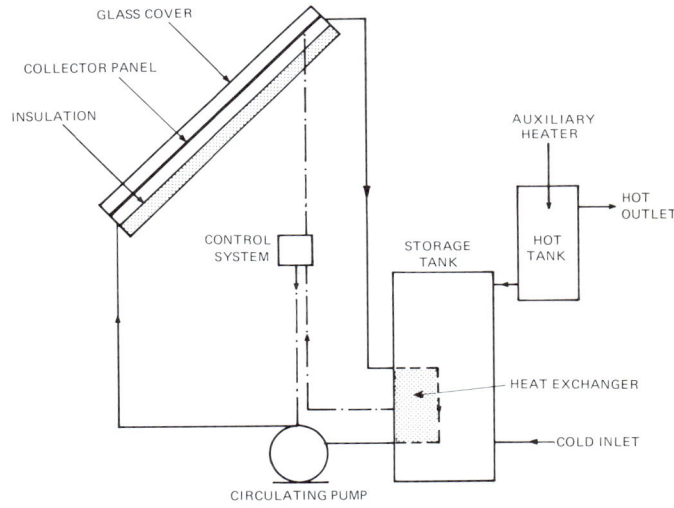

a black plate able to collect radiation which has been diffused by cloud and haze as well as the direct rays from the sun. The heat is transferred into a fluid containing additives to prevent boiling and freezing problems. The heat is then transferred to the storage tank through a heat exchanger. A glass cover over the collector enables the sunlight and diffused heat radiation to pass on to the plate whilst obstructing the outward passage of the radiation deflected back from the collector as it heats up.

As we have said, problems of seasonal and daily variations in solar energy availability render this form of energy impractical as a single source. Nevertheless it can make a useful contribution to overall heating supplies and a growing number of homes are being fitted with such systems. A well designed system can save up to fifty per cent of annual hot water energy consumption but this varies greatly on circumstances. Expenditure on such systems may, however, involve a pay-back period as long as 10 years.

The development of solar heating systems remains in its infancy. Whilst there is every reason to expect the use of these systems to increase and improve in the future, only the more adventurous are likely to be attracted to them at this stage.

Heat pumps

Looking further into the future there may well come a time when the installation of a heat pump for domestic heating and air conditioning will become viable. Units are already being installed in the United States for domestic purposes but work in this country has mainly been centred on developing pumps for industrial and commercial applications.

The basic workings of an elementary heat pump system are explained in Fig. 22. Heat pumps which use the outside air as the heat source and discharge warm air into the building during the winter can also perform the dual function of providing air conditioning in summer. This is simply achieved by a four-way valve which reverses the flow of refrigerant, so that the indoor coil absorbs the heat and the outdoor coil becomes the evaporator.

Heat pumps, however, while offering the prospect of long-term energy savings, remain in their infancy. Whilst a number of suppliers will provide them to order, the lack of any 'off the shelf' keeps their prices high. Current costs are of the order of £1000 for a domestic pump, as against the cost of an equivalent boiler system of around £200.

Fig. 22. How a heat pump works
A heat pump is a device which extracts heat from a cool environment such as the air, ground or water and transfers it to a warmer one; in other words, heat is made to 'go uphill'. This, of course, is exactly what happens in the domestic refrigerator as heat is taken from the ice box and pumped into the room. How is this improbable feat achieved?

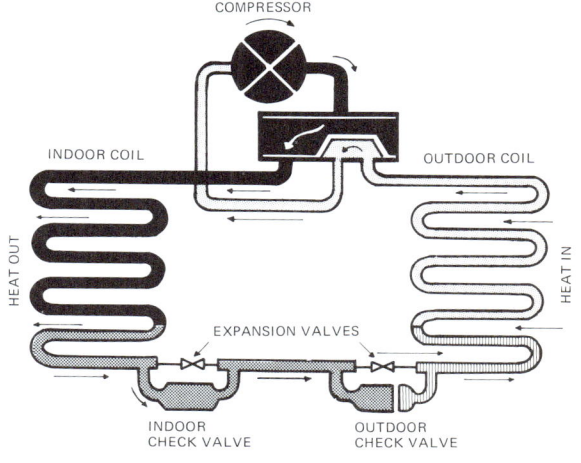

Source: Electricity Council

There are four main steps:
1. *Compression*
   An electrically driven compressor is used to compress a gas (the refrigerant) and in doing so raises its pressure and temperature.

2. *Condensation*
   The compressed hot gas is passed on to the condenser where heat is removed and the gas condenses into a liquid.
3. *Expansion*
   After passing through the condenser the refrigerant is then allowed to pass through an expansion valve where the pressure drops and liquid enters the evaporator at a reduced temperature.
4. *Evaporation*
   In the evaporator the liquid refrigerant evaporates, absorbing heat from the surrounding air.

   The vapour from the evaporator is now returned to the suction side of the compressor where the cycle starts all over again.

   Heat pumps which use the outside air as the heat source and discharge warm air into the building during winter can also perform the dual function of providing air conditioning. This is simply achieved with a four-way valve which reverses the flow of refrigerant so that the outdoor coil now becomes the condenser, and the indoor coil the evaporator.

7. Choosing appliances carefully pays off

Installing a new heating system, or replacing an old appliance, will involve substantial expenditure and will need to take its place amongst the other claims on your household expenditure. But once you have decided to invest in new fuel-burning equipment then it is important to consider the total costs involved.

In arriving at a choice of fuel and heating system it is important to calculate the expected total cost. This will be made up of:

the initial price of the appliance or installation together with the cost of hire purchase or loan repayment if required;

the expected running costs covering the amount of fuel used, standing charges and necessary servicing costs.

In determining the expected running costs of alternative systems you will need to examine not only the comparative prices of the alternative fuels but also their relative efficiencies. (We saw in Table 16 that each fuel performs at different levels of efficiency, depending on the job in hand).

The complexities of arriving at such guideline costs have been recognised by the government, through the Department of Energy, the fuel supply industries and of course, the consumer associations. Leaflets and booklets setting out the comparative costs of alternative fuels are available from the Department of Energy and the British Gas Corporation amongst others.

It must also be borne in mind that in determining running costs it will be necessary to make some assessment of future price rises for the different fuels. Clearly it is impossible to make an accurate forecast on price movements over the entire possible life of the appliance or installation. Indeed, as we all know, fuel prices, both actual and comparative, can change dramatically and at little notice.

It is consequently an area of great uncertainty. What can be said with some assurance, however, is that in the long run the prices of all fuels will keep on rising and probably at rates faster than that of general inflation. Fuel will become relatively more expensive as pressure grows on existing supplies. Many experts anticipate that the real price of our fuels, that is, when the effects of general inflation have been taken into account, will be two or three times the current levels in 20 years' time.

Against this background it is essential to give careful thought to the alternative systems available. We now look at the alternatives offered by the supply industries for central heating systems, room heaters, independent water heating systems and in cooking and other domestic appliances.

Central heating

Central heating systems offer a complete or major supply of heating and hot water for the household. Three fuels – gas, oil and the different types of solid fuel – fire boilers, to perform this dual function. Alternatively, electric central heating systems provide an immersion hot water tank heater and a number of block storage heaters.

As well as separate boiler units, which are normally positioned in the kitchen, back boilers linked to solid fuel or gas fires are available for installation in the living room. Most systems give heat by circulating hot water through radiators. Ducted warm air, supplied by oil, gas or electric heaters, offers an alternative. This normally has a separate appliance built into the installation for water heating. Ducted systems are relatively expensive to install, unless incorporated into the design of the house on construction, but do provide heat more quickly than the radiator systems.

The spread of central heating systems has been rapid.

Fig. 23. The growth in ownership of central heating systems
The popularity of central heating has increased rapidly in
recent years. Eleven million homes are now fitted with such
systems, with well over half of them gas-fired. Nearly all
houses built today include some form of central heating.

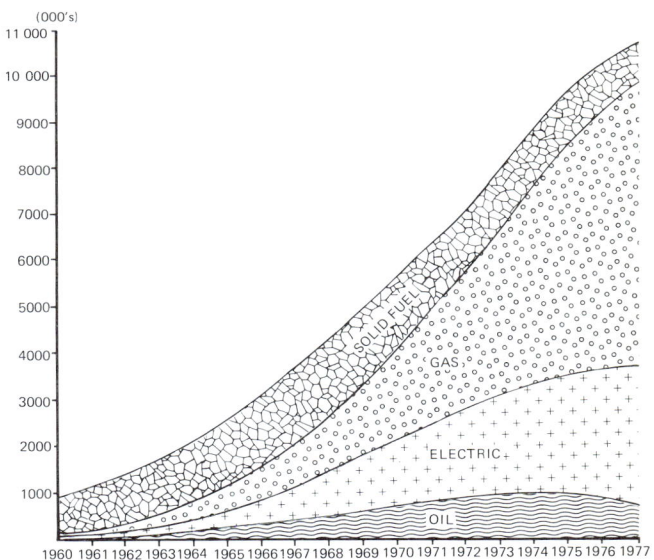

You can see from Fig. 23 that the number installed has
risen tenfold since 1960, counting partial central heating
provided by two or more electric block storage heaters.

The increase in installations, except those which are gas-
fired, has slowed down somewhat in the last few years,
mainly because fewer new houses have been built. Nearly
all new homes built both privately and for local authorities
now include some form of central heating.

There are many types of appliance and installation
available. The main difference between them usually
reflects the size of the house and whether the system is to
offer full or partial heating.

Table 24 sets out the typical costs of alternative full systems suitable for a three-bedroom semi-detached. Prices will vary according to area and supplier as will of course the repayment rates on loan interest. Nevertheless, they should serve as a reasonable guide. Running costs will also vary, of course. So much depends on how much hot water or heat you require and whether these are more or less than the average figures given.

Table 24. Approximate costs for full central heating in a 3-bedroom semi

	Cost of system	Annual payment to finance house for 7 years	Running costs p.a.	Total p.a.
Oil-fired boiler (580 gallons)	£900	£260	£230	£490
Smokeless fuel (60 cwt)	£750	£215	£170	£385
Gas-fired boiler (850 therms)	£750	£215	£160	£375

Figures based on the supply of 200–250 gallons of hot water per week, temperatures of 68°F (20°C) in the living room and 65°F (18°C) elsewhere for a 12-hour day October–April. Prices subject to variation by region.

Note: The figures do not include those for electric central heating because of the difficulties of providing comparisons with hot water systems. Electricity can however provide similar comfort standards – see page 90.

The rapid increase in domestic oil prices coupled with the higher cost of the basic system makes oil-fired central heating relatively unattractive. There is little to choose between the solid fuel and gas alternatives in cost terms although most will value the additional benefits of gas, namely its ease of control, cleanliness and on-tap availability.

The variations in installation and running costs are similar for systems suitable for larger houses, too. For example, the running costs in a four-bedroomed detached house with similar hot water and heat requirements could be:

gas	(1100–1300 therms p.a.)	£220–240 p.a.
oil	(720–740 gals p.a.)	£280–300 p.a.
solid fuel	(90–100 cwt p.a.)	£220–250 p.a.

These variations, whilst significant, are unlikely to warrant the replacement of an existing system with one fired by another fuel. But, where the basic installation is sound, it could well be worth while costing the replacement, say, of an oil-fired boiler by a gas one.

A boiler change, with the installation of a time clock, would cost in the region of £200–300, depending on size required. Costs may be recovered within three to six years. appliance fuelled.

Gas-fired boilers

Most installations are carried out by heating engineeers who are members of CORGI – the Confederation for the Registration of Gas Installers. Members of CORGI can be expected to install gas equipment to the required standards of safety and you should make sure that your installer is CORGI-registered. Those used by Regions of British Gas are all CORGI members, as is British Gas itself, which was instrumental in setting up the Confederation.

Although standard installations are offered with price guidelines, it is normally necessary for the work to be surveyed first. An interesting development has been wall-hung boilers, using either a conventional or a balanced flue to an exterior wall and a small circulating pump which delivers hot water to the tank and radiators. This is particularly valuable where kitchen space is limited. The balanced flue operates by taking in air and returning the products of combustion through a terminal on the outside wall. Regular servicing is recommended by British Gas for all gas-fired boilers. Annual services are carried out on approved installations with customers choosing one of three contract options. British Gas and their installers also provide an 'on request' service.

Oil-fired systems
As we have seen, oil installations are normally more expensive than gas or solid fuel systems. You will have to pay more for the boiler and you also face the additional cost of a storage tank.

However, if you select an oil-fired system, possibly because you live in the country, where other fuels are not so widely available, then it is well worth while investing in a large tank capable of holding 600 gallons or more. Oil distributors normally offer a reduced rate for delivering minimum loads of 500 gallons.

Oil-fired boilers also require regular maintenance. Many distributors offer annual maintenance agreements. Regular servicing is essential to keep the equipment running at peak efficiency and to avoid breakdowns.

Solid fuel systems
Heating engineers are approved by the SFAS, the Solid Fuel Advisory Service, and will undertake surveys and provide quotations. Installation costs are broadly comparable with gas, provided storage space is available. The boiler is slightly cheaper.

Little servicing and maintenance is required, but of course you have to do more work yourself keeping the appliance fuelled.

Electric central heating
A number of systems have been developed to provide electric heating, which used in conjunction with electric immersion heating of hot water, provide full central heating.

The most popular form is the storage radiator, which

stores up heat during the night time, so taking advantage of tariffs with lower kWh rates for units used during this time: see page 67. Heat is stored for release during the day and evening when it is usually needed.

There are about three million systems in use, many of which provide whole house heating with an average of three to five radiators per household.

These radiators provide a good background heat, but cannot immediately be turned up to any degree and therefore are not as flexible in use as other systems. Radiators cost between £50 and £70 each and installation charges are normally £20 per heater. They are available in different sizes and it is important that in any installation the correct size radiators are used in each room. Only minimal use of a radiant fire during the daytime and evening should be made as this is costly, involving the higher rate charges.

Alternatively, fan-assisted radiators can be selected. When extra heat is required the fan can be switched on to draw more heat out from the central store. Prolonged use of the fan may however result in some topping up heating later in the day. Such topping up will be charged at the higher daytime rate.

Although these heaters require special wiring their installation costs are considerably lower than selecting a hot water radiator system. They also have the advantage of providing a flexible system in that further radiators can be added at later dates.

Floorwarming systems provide an alternative. Here the electric heating cables are built into the ground floor of the house and the whole floor radiates heat. Separate arrangements need however to be made for heating upstairs. Electric ceiling heating, which depends on full rate daytime use, is not now being widely installed in new houses.

As mentioned on page 68, an alternative tariff is now available. This is based on a shorter 7-hour period of operation, typically between midnight and 7 a.m., and this offers substantial financial savings over earlier tariffs.

An advantage of all electric heating systems is that they require very little maintenance or attention once installed.

Room heaters

A wide variety of appliances is available using all four principal fuels. A summary of typical appliances and running costs is given in Table 25. You will see that purchase prices vary considerably, but with costs now high you should also examine carefully the likely running costs. Installation costs can also be an important item, especially for electric storage and gas convector heaters.

Table 25. Room heaters: typical purchase prices and running costs

	Appliance cost	Annual running costs	Installation costs
Gas			
Radiant-convector fires	£50–160	£40–65	£10–12
Convector heaters	£40–120	£30–55	£18
Electricity			
Fires	£12–15*	£95–105	–
Block storage heaters	£50–70	£65–70	£20
Solid Fuel			
Open fire	£15–20	£50–60	£5
Closed stoves	£10–80	£40–45	£5–10
Paraffin			
Heater	£40–50	£40–50	–

*Focal point fires are more expensive.

Appliance costs vary according to output, styling, etc. Running cost ranges based on typical living room (10 ft. by 13 ft. with 8 ft. ceiling) maintaining temperature of 68° for 12 hours a day October–April.

Gas fires and convector heaters

The gas fire providing radiant heat from the burners and convected warm air from the top outlets has become the most popular form of room heater. Around 11 million are now in use. In virtually all cases it is ventilated into the fireplace although a 'balanced flue' model is now also available which can be fitted and flued to an outside wall. Running costs will vary depending not only on how much it is used, but also on whether other gas appliances are used in the house.

The installation cost shown is based on a gas point being available by the fireplace. Obviously this will be higher if running extra pipework is necessary. Annual servicing is needed particularly to ensure that the chimney does not become blocked.

Gas convector heaters are becoming popular. They operate on the balanced flue principle, described earlier. Very little maintenance is needed but fixing costs are higher than for fires. They generally involve a separate run of pipe, now at an average cost of £1 a foot.

Electric fires and heaters

Whilst there are nearly as many electric fires in households as gas ones, many of them are of the portable type, used only occasionally. The rapid increase in electricity costs at peak rates, now standing at around 2.8 pence per unit, has been the main reason for the fall in popularity of the fixed hearth type electric fire. Less than a million are now in regular use for full heating.

More popular these days is the use of block storage heaters, described in the preceding section. They have the advantage of operating on the cheaper rate White Meter and can be used individually or as part of a full central heating system. Little maintenance is necessary.

Solid fuel stoves and open fires

The open coal fire was the most popular form of heating before the last war. The Clean Air Act encouraged their replacement by smokeless appliances. But even by then a trend had been established away from this form of heating to other more convenient and labour-saving methods. About 4 million open fires are still in use, these being mainly in rural areas without smoke control regulations. Solid smokeless fuels were manufactured in quantity to meet Clean Air requirements and a variety of these are available.

To improve the efficiency of burning these fuels, closed stoves were developed which are able to control the combustion much better than is possible in an open fire. Consequently it takes only about 18 cwt of smokeless fuel to heat an average living room for a year whilst it can take as much as 28 cwt of coal in an open fire. Smokeless fuels are more expensive than house coal but are still cheaper to use in a closed stove. Again, very little maintenance should be necessary. About a million closed stoves are in use.

Paraffin stoves
Paraffin stoves became popular when oil was cheap in the 1960s. Modern appliances, although much improved on earlier designs, still suffer from the fact that they are flueless. This tends to create moisture. The appliances are also liable to smell if not properly maintained. The burner must be cleaned regularly.

Special care must be taken to minimise fire risks, especially where children are concerned. It is vital to ensure that the heater is positioned away from their reach and properly guarded.

Water heaters

There is a much wider variation in the consumption of hot water between similar households than there is for heating. An average family of four can use as little as 100 gallons of hot water a week or as much as 400–500 gallons. The average is about 250 gallons and it is this figure which has been adopted for the calculations in Table 26.

Table 26. Water heaters: typical purchase prices and running costs

	Appliance cost	Annual running costs
Gas water heaters	£90–110	£40–60
Solid fuel boilers	£150*	£45–65
Electric water heaters	£100*	£60–135

*including water tank

Running costs based on use of 250 gallons of hot water per week at 140°F (60°C) year-round.

Gas water heaters
These are mainly of the instantaneous type, burning gas as the hot water is delivered. They vary in size from small sink units to large 'multipoints'. The costs given in Table 26 are typical for a multipoint installation meeting the full household requirement.

Most now work on the balanced flue principle but there are still over a million older appliances in use either having no flue at all or connected to a separate flue. Regular servicing, at a cost of around £8–10 per year, is essential.

Solid fuel boilers
Many millions of small hot water boilers using coke and other solid fuels have been replaced over the past 10–15 years by boilers for central heating. Most of these were fitted in kitchens but a large number are also of the back boiler type sited behind an open solid fuel fire. They circulate hot water to a hot tank usually in a linen cupboard. About a million or so are still in use but very few are now installed.

For hot water requirements only they consume annually about 15–20 cwt of solid fuel at a current cost of

£45–65 p.a. Very little appliance maintenance is necessary but they do need fairly frequent stoking and cleaning.

Electric water heaters

You can buy separate appliances for sinks or multipoint use or the more popular immersion heaters fitted in the hot water tank. There are around 10 million immersion heaters in use, although about four-fifths are ancillary to other forms of water heating, being used typically in the summer months only.

For full water heating requirements between 4500 and 5000 units of electricity will be used each year, with annual running costs varying between £60 and £135, depending on whether the electricity is charged at off-peak or peak rates. If this is the sole means of water heating it is important to obtain the off-peak rate and lag the cylinder well. The immersion heater and installation without a new cylinder costs £40–50, but double this if a new cylinder is necessary.

Separate electric water heaters are less expensive to install than gas but are much dearer to run if off-peak electricity is not used. Very little maintenance is required, however.

Other home requirements

Finally we look at our remaining heat requirements in the home, namely cooking, claiming 10 pence in every pound we spend on domestic fuel, with lighting and other appliances costing a further 6 pence in the pound between them.

Cooking

Although some solid fuel-fired cookers are still supplied, particularly the range type which also provide hot water, the cooker market is fairly evenly contested today between gas and electricity.

Between 500,000 and 750,000 units of both types are sold each year. Overall there are more gas cookers in use than electric, but electricity predominates in rural areas.

Prices range between about £95 and £300 according to

size and features. Annual fuel use for domestic cookers is typically 78–85 therms for gas and 1500–1600 kWh for electricity. The current cost of providing these fuels varies by area between £40 and £45 a year for an electric cooker. The cost of using a gas fired cooker also depends on what other appliances use the fuel in the home. Annual running costs will vary between £15 and £28 depending on whether the standing charge and the initial higher priced bank of therms – see Chapter 5 – need be charged or whether the lower priced units are appropriate.

Lighting

Electricity is now used for virtually all forms of lighting. Bulbs either of the filament or fluorescent type are marked in watts so that a 100 watt bulb, for example, uses 1 kWh (unit) in 10 hours. Charges per unit (not off-peak) are currently around 2.8 pence, excluding the standing charge. Your lighting needs are unlikely to be claiming much of your fuel costs. Keeping a light burning for security purposes if the house is empty at night should not be considered wasteful. But remember that fluorescent lamps need less electricity than filament ones for the same level of light output.

Other domestic appliances

A wide and growing variety of miscellaneous appliances, almost all electric, are now in use, but though large in number, they normally account for no more than five per cent of total home energy use. Electric refrigerators use about 350–400 units per annum whilst gas refrigerators burn 30–40 therms a year. Home freezers use more energy and consumption may vary acording to capacity and between one model and another. It pays to check when buying.

Laundry appliances vary greatly in use, with washing machines consuming between 1 and 3 units per hour and irons about $\frac{1}{2}$ unit per hour. Dryers also need around 2 units per hour. Radios and televisions consume little electricity, with colour televisions taking rather more than black and white sets at about $\frac{1}{6}$ unit per hour. Toasters use about one unit per hour, and small appliances such as hair dryers and mixers can be expected to run for 3–6 hours for each unit consumed. Vacuum cleaners would operate for 2–4 hours per unit.

Although the purchase prices of these appliances can be expensive, their running costs remain low. So there is little scope for energy saving here, and doing without any of these appliances purely for the sake of energy conservation would be misguided frugality.

Six conclusions on fuels and systems

What then does all this tell us about selecting cost-effective fuels and systems today? And what about the future? Here are six points to bear in mind.

1. No longer should we choose appliances solely on their purchase price. We must also take account of their running costs which have risen rapidly since the energy crisis and are expected to go on doing so.
2. In order to get the best prices it is better to use one fuel for your major energy using appliances than a mixture. For example, if you heat with gas there will be a clear saving if you also cook with this fuel.
3. Considered on price grounds alone, gas and solid fuel offer the most economical forms of central and room heating. The prices of oil and electricity, especially at the peak rate, now make these systems considerably more expensive to run. But if you have a system installed already it is unlikely that it will be worth while changing unless that system uses full-price electricity or the change involved is simply that of the boiler.
4. There are a number of points which should also be considered when you are thinking of buying a new house. Clearly a property with higher insulation standards and a more efficient heating system will entail lower running costs. Such advantages should be expressed in the negotiated purchase price. The

house may be a bit more expensive than one with little insulation and a system expensive to run, but the added investment may be worth it. Conversely, if you invest in cost-effective energy-saving measures you, too, should expect to see some recognition of this in the selling price of your home.
5. In calculating the future running costs of alternative systems you must allow for probable price increases for all fuels. These are likely to be at least in line with inflation in the next few years and certainly so in the mid-1980s and beyond. Supplies of all fuels should be assured, until the end of the century, barring temporary disruption or major political developments. Gas, with its plentiful North Sea supplies and established distribution network, should be well placed to withstand future inflation. Oil also should be in good supply from the North Sea fields. In support, Britain has plenty of coal, the price of which is not so closely governed by world oil prices. The electricity industry will be looking to an increasing nuclear contribution in the years ahead, which should help in limiting future price increases. We will, however, be looking again at likely future price trends in Chapter 9.
6. From an energy-saving standpoint it is attention to the home heating and hot water needs that will bring the greatest benefits. Small electric appliances and lighting, on the other hand, make relatively small energy demands. Other factors such as low maintenance costs and reliability will probably be more important here.

8. Fuel costs need regular checking

Your interest in fuel supplies and prices should not end with a single decision say to install a new system or invest in some form of insulation. The energy markets are frequently changing, both at home and overseas. And these changes may well affect your own position as a fuel buyer. Oil used to be amongst the cheapest of fuels; today it is one of the most expensive. The cost of electricity has also risen dramatically.

At the same time your requirements for fuel may well change. This will obviously be the case if you move house and here you may well find a different type of heating system. New arrivals in the household may also place greater demands on your fuel consumption.

Keeping up to date with tariff and price changes is important; so, too, is understanding you fuel bills. Fuel now represents for virtually every householder a major expenditure item. It is important to budget for these bills within the framework of the household accounts. This question we will tackle in the next chapter, but first let us see what we can do to avoid those nasty surprises when our accounts show unexpected increases.

How to read the meter

If you burn oil or some form of solid fuel and have not entered into a regular supply contract you must keep an eye on how much you use. If you fail to do this and do not reorder in sufficient time you will simply run out and go cold.

But you don't pay for your gas or electricity in this way unless you are one of the minority of customers with slot meters. Electricity and gas are always available at the turn of the tap or flick of the switch. Unless you pay by monthly budget payments it is only afterwards that you will have to face the bill.

Understanding your meter readings will not make the bills smaller but at least you will know how much of the fuel you are using before the bill comes in. It is a simple matter, taking just a few seconds.

Most meters record electricity and gas consumptions on dials although direct numbering mechanisms are gradually being introduced. Dial readings are easily taken by noting down the readings of the four bottom dials from left to right. Where the dial hand rests between two figures it is the lower one which gives the reading (except where the hand rests between 9 and 0). Some dials revolve clockwise and some anti-clockwise. You'll quickly see how yours works.

The progress of meter readings is clearly shown on gas and electricity bills. They both set out clearly the latest and previous meter readings. The difference between the two readings, of course, represents the number of units of electricity or hundreds of cubic feet of gas consumed during the quarter under inspection.

Reading your meter regularly will enable you to check in advance the likely scale of your next bill. It will give warning, for instance, of any particularly high charges associated with abnormally cold weather.

The gas and electricity industries encourage their customers to pay attention to their meter readings. In many cases they are prepared to accept your reading entered on a prepaid card for one or two quarters if it is difficult for their meter reader to gain access. It is also an advantage in checking on those bills you might receive which are based on estimated consumption because the meter reader could not gain access. These estimates are normally based on the corresponding quarter of the previous year. If, however, you consider the estimate to be unreasonable you can notify the supplier of the actual meter reading and get an amended account. Some gas accounts provide an illustration of meter dials for just this purpose and customers are invited to record the positions of the meter hands if they do not accept the estimated reading.

It is then very much in your interest to check your meter readings regularly and keep a track on your own consumption.

Understanding your fuel bill

To convert meter readings to bills charged is a relatively straightforward exercise. Apart from the current unit rates appropriate to each bank of consumption you will also have to take into account the standing charges appropriate to your selected tariffs. Details of these are set out on the bill and in tariff leaflets available from your gas or electricity showroom.

Gas bills

A typical gas bill is reproduced in Fig. 27. The last and previous meter readings are shown on the left-hand side. The difference represents the amount of gas supplied during the quarter as measured in hundreds of cubic feet.

This volume of gas has then to be converted into the amount of heat, or calorific value, in which gas supplies are measured for billing purposes. Gas regions declare the calorific value of the gas they supply and this is printed on the bill. This is the figure, confirmed by government testing, which signifies the heating value of the gas in British thermal units (Btu's – for explanation see page 65) per cubic foot.

To convert to therms (one therm equals 100,000 Btu's) the consumption in hundreds of cubic feet must be multiplied by the declared calorific value and divided by one thousand. As the illustration shows, because the heating value is somewhat higher than 1000 Btu's per cubic foot it means that the consumption in therms will be slightly higher than the consumption in hundreds of cubic feet.

The number of therms is then multiplied by the current price. Under the domestic credit tariff the first 52 therms

NORTH THAMES GAS

A part of the British Gas Corporation

P.O. Box 45, Wembley, Middlesex HA0 1LB

This account is now due for payment. See over for details.

V.A.T. Registration No. 232 177091

ESTIMATED ACCOUNTS

If you consider our estimate to be unreasonable, please complete the DIALS SHOWN ON THE REVERSE of this account and return the account within 7 Days.

Meter Reading: Present (E estimated)	Meter Reading: Previous	Gas Supplied: Cubic feet (Hundreds)	Gas Supplied: Therms	Pence per Therm	Charges £	V.A.T. Rate	V.A.T. Charges £	Total Amount £
9042	8794	248	52.0 204.7	22.8 15.3	43.18	0.000	0.000	43.18
STANDING CHARGE					2.00	0.000	0.000	2.00
						PAYMENT IS NOW DUE		

Account Charged to	Date of Account (Tax Point)	Account Reference for Enquiries	Area No.				Total Amount Due
23 6 78	28 6 78	345246700	1	45.18		0.000	£ 45.18

Mr. P.T. Jones
37, Meadowfield Way
Dagenham
Essex

FOR CUSTOMER'S USE

Date Paid______ Cheque No.______ R

DECLARED CALORIFIC VALUE 1035

16

Fig. 27. Your gas bill
All the information you need to understand your consumption is set out clearly on the bill. Check it carefully and compare it with the bill for the same quarter last year. It will show you whether higher charges are due to increased tariff rates, greater consumption or a combination of both.

per quarter are charged at one rate with all subsequent therms at a lower one. Finally, the standing charge is added in for the quarter, and this makes up the total of your gas bill.

Electricity bills

The total number of units supplied during a quarter is set out on an electricity bill in a similar way to that adopted

Fig. 28. Your electricity bill

The layout of an electricity bill is very similar to that of the gas bill. Meter readings and the consumption figure for the quarter are given on the left-hand side. Consumption may be split into day rate and night rate if you use the White Meter. The controversial fuel cost adjustment charge is shown at the bottom.

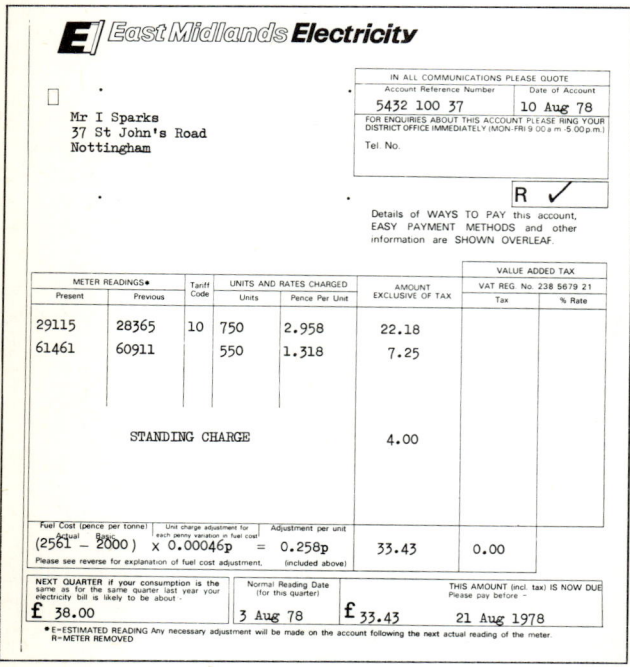

by the gas industry. But this time consumption may be split into two blocks if the customer is using the cheaper rate White Meter for night usage (see page 67 for current tariff rates). The fixed, or standing, charge is also added in to give the total.

When we looked at electricity prices on pages 67 and 68 we noted the existence of the fuel cost adjustment (FCA) charge which is levied by all the Electricity Boards. This represents the additional costs paid by the Central Electricity Generating Board, and passed on to the area boards, for its fuel, mainly coal and oil needed for generating supply.

Introduced at first for industrial customers only, the FCA was extended to cover householders from May 1974. The current rate of FCA, which is incorporated within the unit price, is itemised at the foot of the bill.

The use of this mechanism by the electricity supply authorities automatically to pass on fuel price increases has come under increasing criticism from the Price Commission and consumer protection groups. It means in effect that whilst tariff charges are raised only annually the price of electricity supplied to the customer has been rising every quarter. The FCA may well be dropped in 1979.

Checking household consumption

A useful exercise is to write down the consumption levels of the different fuels you use quarter by quarter. Look back over your old bills to gather together the historic material. When you've established the profile for a full year you may be surprised at the marked seasonal variation. The importance of the winter heating period is basic to all households.

But this profile may also show up some other interesting points. You may be surprised at how much extra electricity you use in the summer months when you switch the hot water load from the central boiler to the immersion heater. Introducing an open fire for winter heating, allowing the central heating boiler to be shut down earlier in the evening, may also have a marked effect. At least, once you have this information you can then act knowledgeably when you use appliances in the future.

Fluctuating temperatures will, of course, affect the amount of fuel you use. By keeping a parallel checklist of average daily air temperatures you will soon see the predominance of the room heating requirement within your total consumption.

Spending a little time on reading your meter and preparing these records will often prevent queries and disputes arising. But some problems are bound to arise with fuel suppliers and this chapter now concludes with some advice on what to do should you feel you have a justifiable complaint.

Questions and disputes on accounts

Gas and electricity accounts are prepared by computers from the information supplied by meter readers. Where access to your home has not been gained it will be necessary for consumption levels to be estimated.

Errors can, of course, occur at either the meter reading or the computer calculation stages although statistically the chances are rare these days. But with more than 100 million gas and electricity bills sent out each year there is bound to be the occasional error.

Meters are tested and stamped by government departments within prescribed limits of accuracy. You are entitled to have your meter tested if you suspect its accuracy, but few are proved to be wrong. If the meter is recording accurately you will have to pay a standard test fee. Bear in mind also that if it is proved that the meter has been undercharging then you will be asked to pay the difference. In the same way of course you will be refunded if it has been overcharging.

Disputes sometimes occur when a house or flat changes occupancy. It is important to have the meters read on leaving and checked on arriving to ensure you pay for what you have used and not for what someone else has.

It is a legal requirement for occupiers to give the supply authorities at least 24 hours notice before premises are vacated. Problems also arise because the householder paying the bill does not appreciate the amount of fuel used. Appliances may have been left on for extra heating because of illness. Understanding the energy facts of life will help avoid disputes of this kind.

Similar disputes do not arise with solid fuel or oil providing the quantities delivered are checked in at the time of delivery. This, however, may cause a problem if for example you have an agreement with an oil supplier to keep your tank topped up. Then if you do not check the amount he is providing and you pay by a standing charge every month you may have a sizeable sum to pay at the end of the year.

Consumer councils

The coal, electricity, and gas industries each have consumer councils appointed by government and consisting of independent representatives chosen from those outside the industry concerned. The Gas Consumers Councils and the Electricity Consultative Councils correspond to the areas of the Gas Regions and the Electricity Boards, and include representatives from the various districts within their areas. There is also a National Electricity Consultative Council to deal with major matters with the Electricity Council, complementing the National Gas Consumers Council set up five years ago when the British Gas Corporation was formed. The Domestic Coal Consumers Council also has branch offices in the sales regions of the NCB.

The Councils are set up to look after customers' interests and make representation on their behalf (a list of their addresses appears at the back of this book). They consider changes in tariffs and charges as well as individual customer difficulties. In certain cases they make representation to government departments as well as the supply industries.

Some criticism has been expressed about the lack of

effectiveness of Consumer Councils, influenced possibly by the lack of publicity for the many cases where customers have been assisted. Also it has been suggested that Councils work too closely with suppliers. But it must be remembered that these Councils can only operate effectively on behalf of customers if they have a close knowledge of the many facets of the particular fuel business in which they work.

The best course of action for you, if you feel you have a complaint or query, is to raise it first with the appropriate department, merchant or supplier direct. Should you not obtain satisfaction here then your next step is to raise the matter with your local consumer council. If the problem concerns solid fuel and you cannot reach agreement with the merchant then you should contact your local office of the Solid Fuel Advisory Service – a list of SFAS regional offices appears at the back of the book.

9. Budgeting wisely avoids shocks

Budgeting for future fuel bills has always been sensible but since the energy crisis it has taken on new importance. The much-increased cost of all fuels has brought greater likelihood of payment difficulties, particularly with the quarterly bills.

The difficulties that have arisen over fuel costs have caused controversy and prompted government to review supplementary benefit heating allowances. There have been reports of special problems with electric heating in some local authority homes and on private estates where the lower installation costs of electric systems has now been more than outmatched by high running costs.

The electricity and gas industries issued a Code of Practice in December 1976. It provides customers with guidance on the steps to take if they expect to have difficulties in paying their bills. It also contains safeguards relating to disconnection for non-payment. Copies of the Code are available at electricity and gas showrooms and offices.

Budget schemes, in existence for many years, have also been given renewed publicity. Savings stamps are now another way of helping to meet these bills. We will be reviewing these various schemes in the following pages of this chapter, concluding with some comments on the likely movements in fuel prices in the years ahead. Clearly these must be of a tentative nature but some 'view' of likely short-term price increases is a fundamental part of the budgeting exercise.

Budget payments

The practice of paying fuel bills as a fixed monthly sum by cash, bankers' order, direct debit or National Giro is growing in popularity. Although such schemes have been operating for many years the gas and electricity authorities are now giving greater publicity to their schemes. Similar arrangements can also be made with many oil and solid fuel distributors.

If you would like to pay your bills by monthly instalment then simply visit your local showroom. The gas and electricity boards will then estimate, by reference to your consumption levels last year, how much you will need to pay each month. And if you've been reading your meter regularly and are armed with the consumption and cost information we've talked about you will be in a better position to judge whether these are reasonable.

Any variations which emerge in terms of the regular outlay and actual consumption will be reconciled periodically. In some cases these will take the form of an additional account or rebate each year, depending on consumption. But practices vary by regional electricity and gas boards and you should enquire about the practice adopted in your area from your local showroom.

You can open a monthly account at any time of the year. But unless you can prove 'hardship' you will have to pay any outstanding account first. For those without bank accounts voucher systems operate in some areas. Here you simply surrender a voucher each time you make a payment at the showroom.

It is of course possible to budget for your fuel bills along with other items of household expenditure through a bank or even a credit card. The fuel industries' own methods, however, are generally more convenient and less costly to use.

Savings stamps

This is another way of helping to spread your fuel costs and is particularly attractive for pensioners, others on low or fixed incomes, or those without a bank or Giro account. Stamps can be purchased in varying nominal amounts from both gas and electricity showrooms.

They can be used against the quarterly costs or alternatively to help pay for appliances or any other services, such as repairs or maintenance. This is a particularly helpful service for those wishing to assist elderly relatives meet their ever-increasing fuel bills.

Prepayment meters

As identified in Chapter 5 you can still elect to have a prepayment coin operated meter for your gas or electricity supplies. But this is not a convenient method for those who have a substantial demand for fuel. And in the case of gas there is the additional problem of relighting the pilot lights when the meter runs out. However, in cases of difficulty and where customers cannot meet their bills in other ways the gas and electricity authorities will agree to provide a pre-payment meter given that it is safe and practical to do so.

Budgeting for inflation

Finally, we look at the problem which has affected us all over the past few years: how can we deal with rising prices? Many times in the previous pages we have noted the importance of estimating and keeping a check on our running costs. To do this we must be realistic about the scale and frequency of future price rises.

Fig. 4, back on page 28, highlighted just how much the price of each fuel has risen since 1972. This period has, of course, been a period of unprecedented inflation and many feel that this is unlikely to be repeated, at least during the next few years. But it would be equally unrealistic to expect prices to steady now for any great length of time.

Central to our problem of inflation has been the massive increases for crude oil demanded by the OPEC countries, raising the price we pay for our domestic oil and placing pressure on the prices of all our other fuels.

However, the current state of the world oil market is depressed. High prices have brought recession and the demand for oil products has dipped. At the same time new areas of supply, such as our own North Sea, Alaska and Mexico, have come on stream, further reducing the demand for OPEC oil.

It is likely then that little in the way of further major price increases can be put through by OPEC. This is especially so given the moderating influence of the leading states of Saudi Arabia and Iran. Rather the establishment of regular, but moderate, increases in price seems more likely until demand picks up substantially.

Where this has been the case, notably in the petrol market, then oil companies have been quick to respond by increasing prices. The growth in demand for petrol in 1978 led to increases in pump prices for all major brands in November.

British Gas was able to freeze tariff prices for the two years up to April 1979. This has proved helpful for customers needing to budget carefully their annual expenditure.

Electricity prices appear to be moving up in line with general inflation. Prices are expected to rise during the year to April 1979, given no major acceleration in their generating fuel costs, by the order of eight to nine per cent. This will be done partly by the five per cent increase in tariffs which took effect from April 1978 and partly by further use of the fuel cost adjustment.

Coal prices will be rising by a similar amount. Although industrial prices were raised on average by ten per cent from the beginning of March 1978, increases for domestic coal were delayed until 1 November, 1978 to provide for a slightly longer summer stocking period this year. But the ten per cent increase introduced at that time means another 15 pence per cwt for house coal and a further 20 pence per cwt for smokeless fuel.

These then have been the performances in price for your fuels over recent months. Of course, they could all change almost at the drop of a hat should the Arabs once more wish to exert their control of world oil prices. And what happens beyond the middle of 1979 cannot be judged with any likelihood of accuracy. Following price and supply developments both in the home market and on the world stage should remain an important part of your continuing interest in fuel. In other words we must 'stay energy conscious', as the title of the following, and concluding, chapter suggests.

10. We must stay energy conscious

We cannot take energy supplies for granted. That is the lesson we have all learnt to our cost since 1973. Not only are we now used to paying prices for our fuel that we would have considered outrageous a few years ago but we also find a major question mark placed against the availability of future supplies.

We have seen that worldwide energy consumption has doubled inside twenty years and will at least double again by the end of the century. And this will come about in the face of dwindling reserves from the established oil and gas fields.

Continued exploration and development will undoubtedly lead to new supplies of all fossil fuels. But these are likely to prove more and more expensive. Oil companies are being forced to search in more remote areas from the Arctic to the Amazon, deep below the sea and further off our coastlines.

Even where oil is still comparatively cheap to produce, as it is in the Middle East, combined action through OPEC has resulted in high prices.

Yet the commercial development of alternative energy supplies remains at a preliminary stage. The nuclear programme has run into major environmental and strategic problems. The task of harnessing the renewable sources like the wind, tides and sunlight is still in its infancy.

All this adds up to energy remaining a seller's market in the long term. And with the current control of oil exports in the hands of OPEC this takes on added significance. The Arab exporters have demonstrated their willingness to use their oil as a political weapon. There is no reason to suppose that they will not do this again.

Britain is of course fortunate amongst her trading

competitors in the developed world. We can look to our North Sea oil and gas reserves, backed by our coal and nuclear industries, to meet all our energy requirements at least until the 1990s and probably for a good deal longer.

But we have seen that we will need to go on charging ourselves the 'market rate' for our fuel during this period. And this means in effect the amount decided by OPEC.

The North Sea will provide us with a useful cushion over the next 10 to 20 years but even during this period we will be unable to isolate ourselves from world prices – nor indeed should we want to. We can expect to receive valuable foreign exchange from our exports in the 1980s. And one day we will almost certainly have to return to the world market, firstly to top up our supplies, and later to replace them.

So there is no room for complacency. Our industries must learn to live with energy as a costly resource and one which is only going to get even dearer. Certainly our competitors in Europe, America and Japan, who are having to face this fact, will become more energy efficient.

In the home also, we have to respond to the high prices of fuel. There is much that we can do and hopefully this book has signposted the areas in which effective action can be taken.

It will certainly pay us to stay energy conscious. Firstly from purely selfish motivation we should get what is going in terms of assistance. A government grant has recently been announced to assist private householders to insulate their lofts. Further incentives to 'save it' could well follow, so keeping up to date with new legislation on grants and incentives should pay off.

Then there is the question of the changing price of each fuel. We saw in Chapter 2 how the price rises for the various fuels have fluctuated; how oil has been transformed from one of the cheapest to one of the most expensive fuels for domestic heating; how also gas has become increasingly the 'best buy' in this area.

Checking the extent of these increases will help you plan your fuel bills. It will also allow you to reach a sensible decision on whether it is worth while to change over to an alternative system of heating or investing in some form of energy saving.

Staying 'energy conscious' also means not relaxing our approach and attitudes to fuel saving. The measures outlined in Chapter 5 are designed to save money without involving any great expenditure. Once you've done them you should think in terms of investing the savings you've made in materials and equipment which will save you even more money (these are set out in Chapter 6). Embarking on such a phased programme will bring major cost savings. Once you've made them they're yours for years ahead.

If the country's lighting bill could be cut by 10 per cent, this would bring a saving of £60 million a year – more than sufficient to pay for the construction of two brand new district hospitals with 850 beds each and full facilities. Yet officials consider this a reasonable target which could be achieved through the better use of existing lights and the installation of more efficient systems.

Tests conducted recently by the Automobile Association in conjunction with the Department of Energy showed that significant savings can also be made in petrol consumption. Eighteen motorists were asked to keep a record for a month of their journeys and the amount of fuel they bought. They were then invited to attend meetings to learn how driving techniques and better vehicle maintenance could improve petrol consumption.

The result was that in the following month the average fuel consumption for these 18 motorists fell by nearly ten per cent. If achieved by all motorists this would bring a saving of 210 million gallons a year, worth about £160 million at pump prices and sufficient for another six district hospitals.

A ten per cent drop in the nation's fuel bill overall would bring a staggering saving of £1,600 million a year, a figure equivalent to the amount we currently spend on our public services and state housing programme.

If such a saving could be achieved and maintained our annual energy consumption by the year 2000 would be cut by the equivalent of 100 million tons of coal. This is just a little less than the National Coal Board's current annual production.

'Saving it' is going to form a central plank of government fuel policy for many years to come. It is in all our interests to make the most of what we have in the North Sea and in our mines.

As these resources deplete in the years ahead, as indeed they must, our attitude to energy and how we use it at home and at work will become critical. The future facing our children is very much in our hands. Energy really is too good to waste.

Appendix: useful addresses

The Gas Consumer Councils
National
National Gas Consumers' Council

5th Floor,
Estate House,
130, Jermyn Street,
London SW1Y 4UL

Regional
Northern

18, Fawcett Street,
Sunderland,
Tyne and Wear SR1 1RH

North Western

Boulton House,
Chorlton Street,
Manchester 1

North Eastern

44, Eastgate,
Leeds LS2 7LR

East Midlands

2, Salisbury Road,
Leicester LE1 7QR

West Midlands

Broadway House,
60, Calthorpe Road,
Five Ways,
Edgbaston,
Birmingham B15 1TH

Wales

St. David's House,
Wood Street,
Cardiff CF1 1ES

Eastern

51, Station Road,
Letchworth,
Herts. SG6 3BQ

North Thames

28, Charing Cross Road,
London WC2H 0DB

South Eastern	Helena House, 348, High Street, Sutton, Surrey SM1 1QA
Southern	2a, Holdenhurst Road, Bournemouth BH8 8AJ
South Western	Royal Building, St. Andrew's Cross, Plymouth, Devon PL1 2OS

Scotland

Scottish Gas Consumer Council	86, George Street, Edinburgh EH2 3BU

The Electricity Consultative Councils

National

Electricity Consumers' Council	5th Floor, 199, Marylebone Road, London, NW1

Regional

London	Room 154, 4, Broad Street Place, Blomfield Street, London EC2M 7HE
South Eastern	1, Boyne Park, Tunbridge Wells, Kent TN4 8EL
Southern	8a, St. Marys Butts, Reading, Berkshire RM1 2LN
South Western	Northernhay House, Northernhay Place, Exeter, Devon EX4 3RL
Eastern	16-18, Princes Street, Ipswich, Suffolk IP1 1QT

East Midlands	Caythorpe Road, Lowdham, Nottingham NG14 7EA
Midlands	Shawton House, 794, Hagley Road West, Oldbury, Warley, West Midlands B68 0PJ
South Wales	Caerwys House, Windsor Place, Cardiff CF1 3UF
Merseyside and North Wales	Martins Bank Buildings, Exchange Street West, Water Street, Liverpool L2 3SP
Yorkshire	Wetherby Road, Scarcroft, Leeds LS14 3HS
North Eastern	Bamburgh House, Market Street, Newcastle upon Tyne NE1 6JD
North Western	Longridge House, Corporation Street, Manchester M4 3AJ

Scotland

Electricity Consultative Council for South of Scotland	North Clyde Street Lane Edinburgh 1
Electricity Consultative Council for North of Scotland	North Clyde Street Lane Edinburgh 1

The Domestic Coal Consumers Council

National	Dean Bradley House 52 Horseferry Road London SW1

The Solid Fuel Advisory Service – Regional Offices

Scottish	Green Park, Green End, Edinburgh EH1 7PZ
Northern	Coal House, Team Valley Trading Estate, Gateshead NE11 0JD
Yorkshire	Consort House, Waterdale, Doncaster, Yorks DN1 3HR
North Western	Anderton House, Newton Road, Lowton, Warrington WA3 2AG
Midlands	Eastwood Hall, Eastwood, Nottingham NG16 3EB
South Wales and West of England	Cambrian Buildings, Mount Stuart Square, Cardiff CF1 1LU
London and Southern	Coal House, Lyon Road, Harrow, Middlesex HA1 2EX

Postscript

The importance of staying "energy conscious" has been further underlined by events since the completion of this book. Developments in December 1978 have once more brought the questions of energy prices and future supplies back into the forefront of public interest.

The scale of the oil price increase agreed by OPEC at Abu Dhabi has jolted the Western world. The increase of $14\frac{1}{2}$ per cent for 1979 is likely to lead to a resurgence in inflation. But perhaps more important has been the decision to move to a system of regularly adjusting prices in the future. Four increases are planned for 1979 alone.

At the same time the disruption in oil supplies from Iran, OPEC's second largest exporter, has served to remind the West of its continuing dependence on imported oil. Little seems to have changed, in terms of our vulnerability to oil shortage, since 1973.

Whilst our growing North Sea supplies should protect us from such a shortage we will still need to charge ourselves the world price for our oil supplies; should our inflation rate accelerate in the coming months, as many now expect, we could find ourselves paying considerably more for all our fuels.

Index

Advanced Gas-Cooled
 Reactors (AGRs), 19
Agrément Board, 77
Appliance costs, 88, 89, 92,
 95, 98
Appliance losses, 54–56, 68–70

Balanced flue heaters, 93
Boilers
 gas-fired, 89
 oil-fired, 90
 solid fuel, 90, 95–96
 wall-hung, 89
Breeder reactors, 19–20
Brent, 41, 48, 49
British Gas Corporation, 42
British Thermal Unit (Btu), 62
Budget payments, 109–110

Calder Hall, 18
Canvey Island terminal, 42
Cavity wall insulation, 74–76
Central heating, 86–92
 gas-fired, 89–90
 oil-fired, 90
 solid fuel, 90
 electric, 90–92
 relative costs, 88
Coal
 see also Solid fuel
 consumption, 34, 36–37
 fuels and heat contents,
  62–63
 markets, 43–44
 prices, 28, 62–63, 112

production, 42–43
reserves, 51
Code of Practice, 109
Commerce, energy
 consumption, 15, 36–37
Confederation for the
 Registration of Gas
 Installers (CORGI), 89
Consumer Councils, 107–108
 addresses, 119–122
Conversion losses, 34–35,
 52–54
Cooking, 96–97
Council grants for insulation,
 74

Disputes on fuel bills, 106–107
Domestic Coal Consumers
 Council, 107
Domestic Credit Gas Tariff, 65
Domestic energy consumption,
 36–38, 54–56
Domestic heating costs, 59
Domestic Prepayment Gas
 Tariff, 66
Double-glazing, 77–78
Draught proofing, 58–60
Ducted warm-air systems, 86

Economy 7 Electricity Tariff,
 68
Ekofisk, 52
Electrical energy, 44

Electricity
  bills, 104–105
  ceiling heating, 91
  central heating, 90–92
  consumption, 36–37, 45
  fires and heaters, 93
  floor heating, 91
  generation, 34–35, 44–45
  meters, 67–68, 101
  prices and tariffs, 28, 67–68, 112
  storage radiators, 90
  water heaters, 96
Electricity Consultative Councils, 107
  addresses, 120–121
Electrons, 44
Energy bank, 17–18, 47
Energy crisis, 23–25
Energy gap, 47
Energy saving, 56, 57–62, 71–84, 115, 117
Energy Think Tank, 22

Flues, 60, 76
Fossil fuel reserves, 17–18
Freezers, 97
French Petroleum Institute, 50
Frigg gas, 41, 48
Fuel bills, 58, 102–107, 109–113
Fuel Cost Adjustment (FCA), 68, 105
Fuel efficiency estimates, 69–70
Fuel oil, 40
Fuel prices, 23, 27–28, 62–68, 90, 96, 112

GNP and energy consumption, 35

Gas
  bills, 102–104
  central heating, 89–90
  consumption, 34, 36–37
  development of North Sea fields, 40–42
  meters, 65–66, 101
  prices and tariffs, 65–67, 112
  room heaters, 93
  water heaters, 95
Gas Consumers Councils, 107
  addresses, 119–120
Gas oil, 40
Geothermal energy, 22
Gigajoules, 62
Good housekeeping, 57–62

Heat losses, 72–73
Heat pumps, 82–84
Hot water savings, 61–62
Hot water tank jackets, 61
Hydro-electric power, 21

Immersion heaters, 96
Industry, energy consumption, 15, 36–37
Insulation, 72–77
International Energy Agency, 31
Iran, 27

JET project, 20

Kerosine, 40
Kilowatt (kW), 67
Kuwait, 27

Lighting, 97
Liquid Natural Gas (LNG), 41
Liquid Petroleum Gas (LPG), 40

Magnox Stations, 19
Manual temperature controls, 78–79
Meters, 65–68, 101, 106
Morecambe Bay gas field, 48

National Coal Board, 43
Natural Gas, *see* Gas
North Sea Gas
  see also Gas
  fields, 48–50
  gathering systems, 48
North Sea Oil
  see also Oil
  development, 38–40
  fields, 48–50
Nuclear fission, 19
Nuclear fusion, 20
Nuclear power, 18–21

Oil
  central heating, 90
  consumption, 34, 36–37
  field recovery rates, 50
  prices, 23, 63–64, 112
  products and heat content, 64
  refining, 39–40
OPEC
  membership, 26
  pricing, 23, 27–32, 112, 114
  production, 25–26
  reserves, 25–26

Paraffin, 40, 94
Petrol, 37–38, 112, 116
Plan for Coal, 43
Plastic film draught proofing, 77–78
Prepayment meters, 66–67, 111
Pressurised Water Reactors, (PWRs), 19

Primary energy consumption, 34, 35
Princeton University, 20

Radiator controls, 61, 80
Radios and televisions, 97
Refrigerators, 97
Renewable energy sources, 21–22
Retail price index, 27
Roof insulation, 72–75
Room heating, 92–93

St Fergus terminal, 48
Saudi Arabia, 27–30
Savings stamps, 110–111
Severn estuary, 21
Solar heating, 22, 80–82
Solar radiation, 21–22
Solid fuel
  see also Coal
  central heating, 90
  stoves and open fires, 93–94
  water heating, 95–96
Solid Fuel Advisory Service (SFAS), 90, 108
  regional addresses, 121–122
Solid wall insulation, 75
South Western Approaches, 50
Standard Domestic Electricity Rate, 67–68
Substitute Natural Gas (SNG), 42

Temperature controls, 78–80
Thermostats, 60, 79–80
Therm, 62, 65
Tidal power, 21
Time controls, 60–61, 79–80
Transport, energy consumption, 36–37

UK energy
  bill, 34
  consumption, 26
  projected supply/demand, 51
  and world energy, 45–46
USA energy consumption, 15–16
United Arab Emirates, 27
Utilisation losses, 54–56, 68–70

Washing machines and dryers, 97

Water heating, 94–96
Wave power, 21
White Meter Electricity Tariff, 67
Wind power, 21
World coal production, 24
World energy consumption, 16, 26
World natural gas production, 25
World oil
  consumption, 16
  shipments, 30